Kant's Aesthetic Cognitivism

Also available from Bloomsbury:

Apperception and Self-Consciousness in Kant and German Idealism, by Dennis Schulting
Freedom after Kant, edited by Joe Saunders
Philosophy, Literature and Understanding, by Jukka Mikkonen
Taste and Experience in Eighteenth-Century British Aesthetics, by Dabney Townsend
The Dialectics of Aesthetic Agency, by Ayon Maharaj

Kant's Aesthetic Cognitivism

On the Value of Art

Mojca Küplen

BLOOMSBURY ACADEMIC
LONDON · NEW YORK · OXFORD · NEW DELHI · SYDNEY

BLOOMSBURY ACADEMIC
Bloomsbury Publishing Plc
50 Bedford Square, London, WC1B 3DP, UK
1385 Broadway, New York, NY 10018, USA
29 Earlsfort Terrace, Dublin 2, Ireland

BLOOMSBURY, BLOOMSBURY ACADEMIC and the Diana logo
are trademarks of Bloomsbury Publishing Plc

First published in Great Britain 2023
This paperback edition published 2025

Copyright © Mojca Küplen, 2023

Mojca Küplen has asserted her right under the Copyright,
Designs and Patents Act, 1988, to be identified as Author of this work.

Series design by Charlotte Daniels
Cover image: Abstract woman's oil portrait
(© Daria Zaseda / Getty Images)

All rights reserved. No part of this publication may be reproduced or transmitted in any form or by any means, electronic or mechanical, including photocopying, recording, or any information storage or retrieval system, without prior permission in writing from the publishers.

Bloomsbury Publishing Plc does not have any control over, or responsibility for, any third-party websites referred to or in this book. All internet addresses given in this book were correct at the time of going to press. The author and publisher regret any inconvenience caused if addresses have changed or sites have ceased to exist, but can accept no responsibility for any such changes.

A catalogue record for this book is available from the British Library.

A catalog record for this book is available from the Library of Congress.

ISBN: HB: 978-1-3502-8951-2
       PB: 978-1-3502-8955-0
       ePDF: 978-1-3502-8952-9
       eBook: 978-1-3502-8953-6

Typeset by Integra Software Services Pvt. Ltd.

To find out more about our authors and books visit www.bloomsbury.com
and sign up for our newsletters.

To my parents
for always loving and supporting me
And to my two little stars, Blaž and Iva

Contents

Abbreviations of Kant's Works   viii

Introduction   1

1   Aesthetic cognitivism in the arts   11
2   Kant and art as expression of aesthetic ideas   47
3   Artistic expression of aesthetic ideas and therapeutic self-knowledge   75
4   Cognitive value of representational and non-representational art   97
5   The aesthetic thesis of Kant's cognitivism   121
6   Kant and aesthetic cognition   143

Conclusion   165

Notes   171
Bibliography   182
Index   196

Abbreviations of Kant's Works

References to Immanuel Kant are given in the text to the volume and page number of the standard German edition of his collected works: *Kants gesammelte Schriften* (KGS), herausgegeben von der Deutschen Akademie der Wissenschaften, 29 vols. (Berlin: Walter de Gruyter, 1902). References to the *Critique of Pure Reason* are to the standard A and B pagination of the first and second editions. Specific works cited are referred to by means of the abbreviations listed below. The translations used are also listed below and, except in the case of the *Critique of Pure Reason*, are referred to immediately following the reference to the volume and page of the German text.

A/B	*Kritik der reinen Vernunft* (KGS 3–4). *Critique of Pure Reason*, trans. W. S. Pluhar, Indianapolis: Hackett Publishing Company, 1996.
Anthro	*Anthropologie in pragmatischer Hinsicht* (KGS 7). *Anthropology from a Pragmatic Point of View*, trans. R. B. Louden, Cambridge: Cambridge University Press, 2006.
FI	*Erste Einleitung in die Kritik der Urteilskraft* (KGS 20). *First Introduction to the Critique of the Power of Judgment*, trans. P. Guyer and E. Matthews, Cambridge: Cambridge University Press, 2000.
KU	*Kritik der Urteilskraft* (KGS 5). *Critique of the Power of Judgment*, trans. P. Guyer and E. Matthews, Cambridge: Cambridge University Press, 2000.
V-Anth/Fried	*Anthropologie Friedländer* (KGS 25). *Anthropology Friedländer*, trans. R. R. Clewis and G. F. Munzel, in *Lectures on Anthropology*, R. B. Louden and A. W. Wood (eds), Cambridge: Cambridge University Press, 2012.
V-Anth/Mron	*Anthropologie Mrongovius* (KGS 25). *Anthropology Mrongovius*, trans. R. R. Clewis and G. F. Munzel, in *Lectures on Anthropology*, R. B. Louden and A. W. Wood (eds), Cambridge: Cambridge University Press, 2012.

Introduction

The idea that works of art can serve as a source of knowledge does not have well-established grounds in the history of aesthetics and philosophy of art. Works of art have been traditionally considered as having an aesthetic and ornamental function, rather than an epistemic one. On the other hand, the fact that the arts have an important status and function in our cultural and educational practices reinforces the idea that works of art play a significant role in enriching human understanding. However, while for most of us it seems quite obvious that art has a cognitive and educational function, it is less clear in what exactly this function consists of. What kind of knowledge can we acquire from works of art? How artworks qua artworks can be a source of knowledge, considering that a great deal of them include statements, description of characters and situations that are completely fictional. Thus, the main worry is how can works of art as fictional entities give us knowledge about the world? Can they give us knowledge that is non-trivial (not known before the work appears), warranted (confirmed by the work) and unique (that cannot be obtained by other means)? Furthermore, if artworks can deliver such knowledge, is this knowledge relevant for their aesthetic or artistic value?[1] That is, are works of art good or bad partly in virtue of the knowledge they give us or fail to give us respectively?

These are some of the questions that have in recent years permeated the debate in aesthetics and philosophy of art. On the one hand, aesthetic cognitivism argues that artworks are an important source of knowledge and that such knowledge (partly) determines their aesthetic value.[2] That is, aesthetic cognitivism is construed as a conjunction of two claims: (i) epistemic claim stating that artworks give us knowledge or some form of cognitive insight about the world and (ii) aesthetic claim maintaining that the knowledge thus obtained is relevant for the aesthetic value of an artwork (Gaut 2006). It is partly because artworks advance knowledge that they are considered as good works of art.[3]

On the other hand, aesthetic anti-cognitivism denies that art can give us any knowledge, at least knowledge that is non-trivial, warranted, unique and aesthetically relevant. That is, aesthetic anti-cognitivism can take two forms: (i) extreme aesthetic anti-cognitivism, which denies both the epistemic and aesthetic claims of aesthetic cognitivism. The well-known advocates of this position are Jerome Stolnitz (1992) and T. J. Diffey (1995), both of them arguing against the idea that works of art possess any cognitive merits, and (ii) moderate aesthetic anti-cognitivism, which holds the epistemic claim, yet denies the aesthetic claim. The main proponents of this position are Peter Lamarque and Stein Olsen (1994), who claim that artworks have the ability to provide different types of knowledge about the world and human experiences, yet that such knowledge is never aesthetically relevant. That is, cognitive merits of artworks do not constitute their aesthetic merits.

Moderate aesthetic anti-cognitivism is currently a prevalent anti-cognitivist view that stands in opposition to aesthetic cognitivism. While both of them maintain that artworks have an important cognitive function in terms of advancing some type of knowledge, they disagree on the question concerning the aesthetic relevance of such function. According to moderate anti-cognitivists, cognitive function of artworks does not matter aesthetically, whereas aesthetic cognitivists claim that it does.

Aesthetic cognitivists support their position by means of two main arguments. First, they argue that our critical vocabulary of aesthetic appraisal is often imbued with cognitive-value terms: 'We praise works for their profundity, for being psychologically penetrating, for giving an insightful perspective on the world. We decry them for being shallow, distorted, inane or full of worn clichés' (Gaut 2006: 167).[4] But if our aesthetic evaluation of artworks has an inherent cognitive component, then this indicates that cognitive function of artworks is aesthetically relevant. Works of art are good (partly) in virtue of possessing cognitive value. Second, they claim that artwork's advancement of true or false thematic statements determines their aesthetic merit or demerit. In support of their position, they ask us to perform a following thought experiment:

> Imagine an equally successful novel as Our Mutual Friend, which had as its theme the contrary of that novel's theme. This novel – call it Our Common Enemy – has as its theme that the worship of money is the greatest of human goods, and other values such as love, companionship and human achievement are mere glittering illusions, baubles to be cast aside for the enduring value of wealth.
>
> (Gaut 2007: 125)

If we can imagine the possibility of such a scenario without encountering any incongruities in the particular development of such theme, then the cognitive value of thematic truthfulness does not matter aesthetically. Yet, we cannot entertain such imaginative possibility for

> as we try to develop the theme through the details of characterization, narrative and how the reader is invited to respond to events, we end up with patent absurdities. [...] It is precisely in developing a theme through the details of a work, then, that its truth becomes of artistic relevance, in part because it allows a depth and correctness of characterization that the false theme resists.
>
> (Gaut 2007: 183)

Hence, they conclude that cognitive merit (or demerit) of the work, such as the truth (or falseness) of the thematic statement, is aesthetically relevant.[5]

Moderate aesthetic anti-cognitivists accept cognitivists' claim that artworks have a substantial cognitive value; furthermore, that profound and reflective nature of thematic statements as expressed in the narrative details of the work is part of artwork's aesthetic value. Yet, they deny that truthfulness of thematic statements matters aesthetically. While aesthetic anti-cognitivists agree that being profound, reflective and insightful is an important part of our critical vocabulary, they deny that these terms possess a genuine cognitive value: 'The point is that we can acknowledge the power and seriousness of literature, even its "cognitive" nature, without supposing that its seriousness (value) lies in its ability to advance knowledge' (Lamarque 2010: 80). A work can be serious, reflective and profound without offering truth and knowledge: 'The mistake is to suppose that to be serious or reflective a work must in effect teach something. Yet there is no such implication. A work is serious if it treats of a serious subject matter. But it can do that without being true and without presenting a view that ought to be endorsed?' (Lamarque 2009: 258). An artwork's expression of profundity and insightfulness is an aesthetic, rather than cognitive property, for it is a result 'not so much that it implies a true proposition, but that it can be interpreted as about humanly interesting concerns' (Lamarque and Olsen 1994: 329–30) developed in an original and powerful manner: 'It is usually a work's treatment of a theme which is judged profound or sentimental and that is a mark of originality, lack of cliché, attention to detail, and so forth, rather than truth' (Lamarque 2006: 130).

Moderate aesthetic anti-cognitivists typically support their position by offering the following two arguments. First, truthfulness of cognitive claims is never a component of critical discourse and there is 'an absence of argument about whether or not a particular proposition or set of propositions implied in a literary work is true or false. Indeed, critical work is ended when it has been demonstrated how such a general proposition or set of propositions organizes the various features of a work into a meaningful pattern' (Lamarque and Olsen 1994: 332). What matters aesthetically is not the truthfulness of work's thematic claims, but rather exploration of the theme as manifested in the narrative details of a work, particular description of characters and emotional force of the artist's style. Cognitive and aesthetic appreciations, anti-cognitivists argue, are substantially different for they invite distinct modes of attention due to their different subjects of appreciation. Cognitive appreciation is an appreciation of work's truth and knowledge it offers and as such it attends to its 'reasoning, its persuasiveness, its arguments, and its goal of truth' (Lamarque 2006: 134). Aesthetic appreciation, on the other hand, is an appreciation of the 'design or structure of the work, how its parts cohere into a satisfying aesthetic whole' as well as of the 'broader thematic characterization that both unites discrete elements of the subject matter and offers a filter through which to reflect on it' thereby inviting 'different perspectives on their content, different ways in which the subject matter can be conceived' (Lamarque 2006: 133–4). What matters from the aesthetic point of view is not the theme itself and its relation to the external world (that is, whether the thematic statement agrees or disagrees with the facts of the world), but rather how the thematic statement, irrespective of its truth value, coheres with the narrative details of the work and how convincingly it is supported by the imaginative particularities of the subject matter.

Second, artworks differ considerably in the cognitive claims they advance, which can often contradict each other; yet, they can be equally aesthetically successful. If this is the case, then cognitive value of thematic truthfulness does not matter aesthetically: 'No doubt a different artistic treatment could present a theme of equal interest albeit formulated in a proposition which is the precisely negation of this one. It is the content of the proposition, what it is about, not its truth as such, that confers interest' (Lamarque and Olsen 1994: 330). Works of art have cognitive merit (or demerit) when the thematic statement agrees (or disagrees) with the facts of the world and they have aesthetic merit (or demerit) when the thematic statement shows coherency (or incoherency) with the particularities of the work. Furthermore, given that works of art essentially deal with thematic statements that are of deep philosophical, social and psychological interest, and that such statements 'are rarely demonstrably true or false' (Lamarque 2006: 137), this implies the aesthetic irrelevance of their truth value.

The truth value of thematic statements is not part of our aesthetic appreciation of artworks. The purpose of art, aesthetic anti-cognitivism argues, lies in the imaginative realization of the theme, rather than in the theme itself and what it communicates. That is, what matters in art is the organization and structure of a work's formal properties, subject matter and thematic characterization, and how all these elements cohere into a unified pattern, thereby occasioning a certain kind of aesthetic enjoyment.

Both aesthetic cognitivism and moderate aesthetic anti-cognitivism have their own merits. Aesthetic cognitivism is right in claiming that there is much more to an artwork than just being aesthetically pleasing. We often admire artworks for their insightfulness, while we criticize other works for being shallow and superficial. Thus, it appears that our vocabulary of artistic appraisal is charged with cognitive value terms. While these terms might not have necessary connection to knowledge typically defined as justified true belief (propositional knowledge), they can nevertheless indicate some form of cognitive advancement that does not necessarily require evidential and argumentative warrant. For example, various proposals have been made that works of art contribute to the refinement of our conceptual vocabulary, promote emotional education or experiential knowledge such as learning what it is like to be in a certain situation or a kind of person, or provide axiological understanding, which involves integrating factual knowledge with emotional, social and psychological dimension. Thus, even though, as anti-cognitivists claim, being profound and insightful is not a matter of truth and factual knowledge, but rather an aesthetic property of the manner with which the thematic claim is developed and explored, namely in an original way and with attention to narrative details, it does not follow from this that such aesthetic property fails to promote some form of cognitive advancement. After all, if aesthetic treatment of a theme displays originality, meaning that it is not determined by any known rules, then this suggests that it must afford a novel and unique cognitive experience, since any treatment of a theme that is governed by known rules must be to that extent imitative and trivial. Thus, original exploration of a thematic statement, irrespective of its truth value, must add to our stock of knowledge in some way, for example, by means of offering a fresh perspective or a new understanding on the thematic statement that we did not consider before. Similarly, if a presentation of a thematic statement demonstrates the aesthetic quality of attention to details, then it can be cognitively relevant by way of offering a more nuanced and rich perspective on the theme, thereby deepening our understanding of it in different ways. As Elvio Baccarini and Urban Milica express a somewhat similar idea, 'the artwork, although not directly offering truths, advances our knowledge through challenges and, thus, through offering improvements to what we believed earlier' (2013: 482). Such improvements are all the more cognitively relevant if, as anti-cognitivists claim, works of art deal with thematic statements that are of perennial interest and can be seldom demonstrated as true or false. For it is only through a collision of competing statements that we can hope to arrive at a more complete understanding of each of them.

Yet, this is not a particular quality of artworks alone, aesthetic anti-cognitivists claim, and in this respect, they make a good point. The kind of knowledge that cognitivism claims art is supposed to give is something that is either already known or can be obtained by other means (through science, philosophy or real-life experiences), even more effectively so. But if knowledge can be obtained by non-artistic means, then what is so special about the cognitive value of art?

One way to defend the position that art has a unique and aesthetically relevant cognitive value depends on showing that aesthetic value, essential to artworks, is cognitive. This is a difficult task to begin with, considering that aesthetic value has traditionally been distinguished from cognitive value based on the view that aesthetic experience depends on the feeling of pleasure or displeasure, and that such feelings are essentially non-cognitive.

The aim of my book is to express a critique of this view and to show that aesthetic experience of the beautiful (and ugly) is a kind of cognitive experience, in particular the kind of experience that reveals those aspects of the world that we cannot fully obtain through ordinary cognition. In short, I hold the view that the value of art lies in aesthetic value, that is, in facilitating aesthetic experience of pleasure and displeasure due to the aesthetic form, whereby the elements that comprise this aesthetic form may not be merely perceptual elements, but semantic or intellectual elements as well. Yet, in contrast to other aesthetic theories of art, I hold that a distinctive cognitive value is also an essential feature of art. I aim to reconcile these two seemingly incompatible positions by arguing that aesthetic value is a species of cognitive value. Accordingly, my approach substantially differs from the approach typically taken by the defenders of aesthetic cognitivism, since they take it as a standard view that aesthetic value is significantly different from cognitive value, yet that on certain occasions they can influence each other. However, how this interaction between aesthetic and cognitive value takes place is a matter of controversy and the main target of objections given by aesthetic anti-cognitivism. No such objections arise if we take aesthetic value as a kind of cognitive value. I develop my proposal in light of Kant's theory of artistic expression of aesthetic ideas as put forward in the *Critique of the Power of Judgment*.

I begin with a critical examination of the main approaches given to defend the epistemic claim of aesthetic cognitivism, focusing on the propositional, conceptual, experiential knowledge approach and John Gibson's account of axiological understanding. I argue that none of these approaches are fully successful either because they fail to meet the main objections given by anti-cognitivists (i.e. cognitive triviality, no-justification, textual-constraint and aesthetic-relevance objections) or because they fail to present the form of epistemic thesis that counts as a genuine cognitive achievement. In conclusion, I suggest that the cognitive value of artworks is best construed in terms of the epistemic notion of objectual understanding as developed by contemporary epistemologists. After considering few current attempts of applying objectual understanding to works of art, I propose the idea that works of art offer (objectual) understanding of abstract phenomena in that they reveal various relationships between introspective, emotional and affective properties that, as suggested by contemporary studies in cognitive science, constitute the semantic content of abstract concepts. I develop my approach in light of Kant's theory of artworks as expressions of aesthetic ideas.

In the second chapter, I take a closer look at Kant's theory of aesthetic ideas. I argue that artistic expression of aesthetic ideas serves as indirect presentation of the conceptually indeterminate semantic material that constitutes the content of abstract phenomena. That is to say, aesthetic ideas make various introspective, emotional, affective and other experience-related properties associated with our abstract concepts (such as concepts of time, justice, alienation, hopelessness, grief, death, truth, etc.) salient by connecting them with particular imaginable representations and thereby make them more cognitively accessible to us. They provide us with an additional source of information that is required for a more complete understanding of abstract phenomena. Accordingly, my account offers an insight on problems related to the apprehension of meaning and cognitive processing of abstract representations that have been of interest to cognitive science. This is a problem about the question of how abstract concepts are represented considering they do not have bounded, identifiable and clearly perceivable referents as concrete concepts have. Rather, they are grounded in situations, introspection and emotions and are accordingly much more difficult to understand than concrete or determinate concepts. The gist of my argument is that aesthetic ideas afford us with an opportunity to imagine the introspective, emotional and affective aspect connected with our abstract concepts, thereby imbuing them with a more substantive meaning and understanding.

In Chapter 3, I integrate my account of aesthetic ideas with contemporary theories of self-knowledge in order to show how artistic expression of aesthetic ideas can promote the kind of self-knowledge that leads to self-development and self-change, namely therapeutic self-knowledge. As it has been argued recently by psychologists and philosophers of mind, to obtain therapeutic self-knowledge we must relate to our own mental states from a first-person perspective, since it is this perspective that necessitates the feeling of ownership and authority over our own mental life and can thus facilitate self-change. Yet, we must also regulate first-person perspective by adopting a more distant, third-person perspective on our own self, since it is this perspective that allows us to acknowledge the meaning of our experiences in the larger context of our life. I argue that artworks as expressions of aesthetic ideas allow us to experience such dual perspectives, thereby helping us to make sense of our own internal experiences as well as allowing us to recognize the meaning of these experiences in the larger context of our lives.

In Chapter 4, I examine the relationship between various artforms, such as literary, visual and musical art and their respective cognitive effects in terms of expressing aesthetic ideas and thereby promoting objectual understanding. In order to resolve certain discrepancies within Kant's text, specifically his inconsistent view on the nature of the relationship between non-representation art and their ability to express aesthetic ideas, I propose a distinction between two kinds of aesthetic ideas, namely, productive and reproductive aesthetic ideas communicated by artistic form and mere sensations (mere tones and colours) respectively. I conclude that in contrast to representational art, non-representational art (such as absolute music and visual abstract art) expresses reproductive aesthetic ideas by means of already-existing associations and as such their cognitive value in terms of promoting objectual understanding is significantly limited.

In the fifth chapter I offer a new defence of the aesthetic thesis of aesthetic cognitivism in light of Kant's theory of artistic beauty. I demonstrate that Kant holds an extreme form of aesthetic cognitivism in respect to art, namely, the view that all works of art must exhibit expressive and hence cognitive value in order to be aesthetically successful. I propose the connection between artwork's cognitive and aesthetic value by referring to Kant's notion of taste (that is, experience of free harmony), which is responsible not merely for occasioning the feeling of pleasure in the beautiful but also for providing a coherent meaning and sense to the manifold.

In the sixth and final chapter, I provide a detailed theoretical background for my interpretation of Kant's aesthetic cognitivism. I argue that aesthetic value is a species of cognitive value in the general sense of cognition as the purposive ordering of experiences into meaningful wholes. I demonstrate that the aesthetic feeling of pleasure in the experience of the beautiful has a cognitive function similar to that of a determinate concept in ordinary cognition in that it serves a role not only of recognizing a relationship between cognitive faculties of imagination and understanding as freely harmonious, but also as a felt sense that guides the imaginative ordering of the manifold and making free harmony possible.

1

Aesthetic cognitivism in the arts

One of the main debates in contemporary aesthetics and philosophy of art concerns the question whether we can learn anything from art. On the one hand, aesthetic cognitivism argues that works of art serve as an important source of knowledge about the world (epistemic claim) and that such knowledge we thus gain contributes to the aesthetic value of an artwork (aesthetic claim). That is, we appreciate works of art partly in virtue of knowledge they give us. On the other hand, aesthetic anti-cognitivism denies one or both of these claims. While extreme aesthetic anti-cognitivism denies both the epistemic and aesthetic claims arguing thereby against the idea of works of art serving as a source of any kind of knowledge, moderate aesthetic anti-cognitivism, on the other hand, holds the epistemic claim and denies the aesthetic claim. That is, it argues that artworks can give us different kind of knowledge about the world; yet, that such knowledge is never aesthetically relevant. Cognitive merits of works of art never constitute their aesthetic merits.

My aim in this chapter is to examine and re-evaluate main aesthetic cognitivists' approaches in light of anti-cognitivists objections. Given that the number of these approaches is not insignificant, I classify them into four broad kinds of approaches: propositional knowledge approach, experiential or subjective knowledge approach, conceptual or philosophical knowledge approach and neo-cognitive approach, which stresses the cognitive value of understanding, rather than knowledge, in respect to art. I raise a number of questions about the adequacy of these approaches, in particular, whether they can meet all four main objections given by aesthetic anti-cognitivists. First, the cognitive triviality or banality objection, which claims that knowledge acquired from works of art must not merely entertain truisms, but rather offer new and interesting cognitive insights about the world and human experiences. Second, no-justification objection maintaining that in order for works of art to deliver knowledge they must also provide evidential or argumentative support for the conveyed cognitive claims. Third, the textual-constraint objection, which states that any cognitive value possessed by works of art must be properly part of the content of the work. That is, artworks should not be only used instrumentally to promote various cognitive skills; rather, the artwork itself in virtue of its narrative details must display cognitive insights. Fourth, the aesthetic-relevance objection, namely, the idea that artwork's cognitive value must (partly) constitute its aesthetic value. That is, cognitive value must be aesthetically relevant in the sense that it makes the artwork aesthetically better. I will argue that none of the approaches are fully successful either because they fail to meet all four objections given by aesthetic anti-cognitivists or because they fail to offer a form of cognitivism that counts as a genuine cognitive achievement. In conclusion, I explore neo-cognitivist proposal that the epistemic notion of objectual understanding, rather than the concept of knowledge, can more properly account for the cognitive value of art. I further develop such an approach in light of Kant's theory of aesthetic ideas.

Propositional knowledge approach

The most straightforward way to defend aesthetic cognitivism is to claim that artworks give us propositional knowledge or knowledge of facts (i.e. knowledge that something is the case). This is the kind of knowledge that is defined in the conventional sense as a justified true belief. Presumably, works of art can serve as a source of propositional knowledge given that authors often incorporate factual truths about the real world into their work. This way we can learn, for example, many facts about the intricate Japanese work ethics by reading Amelie Nothomb's satirical novel *Fear and Trembling* (1999), which portrays adventures of a Belgian young woman going to Japan to spend a year working at the Yumimoto Corporation. Or, we can obtain information about the early medical history by watching television show *The Knick* (2014), which follows Dr John W. Thackery and the hospital staff working in the Knickerbocker Hospital in New York during the early twentieth century.

While the idea of acquiring propositional knowledge from works of art appears to be intuitively true, it is, however, more difficult to answer the question as to how exactly do we acquire factual information about the real world from works of art given that such works are not meant to be taken as being about the real world. A great deal of artworks that we consider cognitively valuable are fictional and as such referring to the imaginary or make-believe characters and state of affairs. Fictional artworks, as Peter Lamarque and Stein Olsen point out, instantiate fictional statements distinctive for their lack of truth values – they can neither be true or false since the things they refer to are non-existent (1994: 31–2). But if works of art do not purport to describe the real world, but rather stand 'outside the polar relation between truth and falsehood' (Konstan 2015: 9), then we have no rationale to claim that they can serve as a source of propositional knowledge. This approach is accordingly faced with the following paradox of fiction[1]:

1. Artworks can promote propositional knowledge about the world insofar as they refer to the real-world state of affairs.
2. However, artworks are fictional, i.e. the statements, characters and situations described or presented in works of art are imaginary and true for the fictional world alone.
3. Thus, artworks cannot give us propositional knowledge about the real world.
4. Yet, it is intuitively true that many artworks have cognitive value in the sense of promoting propositional knowledge and stating true beliefs about the real world.
5. Thus, how can something that is fictional and not true of the real world nevertheless give us propositional knowledge about the real world?

In response to the paradox, different solutions have been proposed. One such solution, given by Stacie Friend (2007), is simply to deny the premise two of the argument and to point out variety of artworks that are non-fictional. Examples are portraits, documentaries and autobiographical works that refer to actual state of affairs, portray real people and can therefore convey true beliefs about the real world. While this might be one way of securing aesthetic cognitivism, it is not particularly successful for it fails to account for the great deal of artworks that are fictional, yet which we consider as cognitively valuable in virtue of their advancement of factual information about the world. A different solution is proposed by I. A. Richards (2004) who denies the existence of the paradox by claiming that while artworks can often evoke in us feelings of deep importance and significance, this is a mere feeling of emotional, rather than cognitive significance. Works of art can deliver intense emotional beliefs or belief-feelings, which we often confuse with scientific beliefs. However, the most widely discussed and plausible solution is given by the revisionist account of the propositional knowledge approach, which makes use of the distinction between artwork's subject level (that is, content or plot) and its thematic level. While the artwork's subject level cannot serve as a source of propositional knowledge about the world for it is fictional (i.e. construed from fictional characters and state of affairs), the artwork's thematic level, which is inferred from the content of the work, can give rise to general thematic statements about the human world and these statements can be assessed as true or false (Weitz 1943, Kivy 1997, New 1999, Mikkonen 2013). According to this proposal, cognitive value of artworks consists in delivering true thematic statements, in particular statements of social, philosophical, metaphysical and psychological kind. These statements can occur either explicitly in the content of the work or implicitly, in which case they are extracted from the content of the work by the audience. In order to illustrate the distinction between the artwork's subject and thematic level, I will consider Edward Albee's play *Who's Is Afraid of Virginia Wolf* (1962).

The play tells a story about a volatile marriage of a middle-age couple named George and Martha. After they return home drunk from a faculty party, Martha reveals she has invited a young married couple Nick and Honey, whom she had met at the party, for a drink. As the guests arrive, a long night of cruel games, humiliations, verbal abuse and painful confrontation begins, where George and Martha reveal their painful secrets and expose their illusions. The subject level of the play is wholly fictional – that there is a couple, named Martha and George, and that they are returning from the faculty party are all fictional elements that have no reference beyond the work itself. They are fictional truths, which can be justified by pointing out to the particular scene in the play. Nevertheless, the play also contains various implicit thematic statements, which can be revealed by peeking beneath the story or by considering what Morris Weitz calls 'the depth meaning' or the 'second order meaning' (1943: 344). For instance, we can see that behind insults and horrifying behaviour between George and Marta, they still have love for each other, but which has, due to George's academic failure and loss of a child, converted into hate. There is then an implicit thematic statement, namely, that love (not merely love between Martha and George, but love in general) can quickly transform into hatred. In addition, since George and Martha connect to each other best when trading insults, another implicit thematic statement offered by the work is that marriage (not only marriage between George and Martha, but marriage in general) can often lead to a series of games, lies and fantasies that couples play with each other in order to hide their true emotions and to cope with life easier. These are the kind of implicit thematic statements that have reference to the real world and can thus be evaluated as true or false.

On the face of it, this approach gives a reasonable solution to the paradox of fiction. By making a distinction between the artwork's subject and thematic level, it allows for the possibility of acquiring true beliefs from works of art in spite of their fictional nature. General thematic statements can be revelatory of the real world, specifically the world of human experiences. Nonetheless, this approach is not fully satisfactory as it fails to meet all four objections given by anti-cognitivists. First, it has been argued, most notably by Jerome Stolnitz, that thematic statements conveyed by works of art do not advance new factual beliefs, but merely entertain truths we already know: 'Only rarely does an artistic truth point to a genuine advance in knowledge. Artistic truths are, preponderantly, distinctly banal. [...] These are the slight, dull, obvious realities which have been obscured by the grandiose pieties of cognitivism' (Stolnitz 1992: 200). We know before encountering the work that, say, 'love can quickly transform into hate' or that 'people often hide their true emotions behind insults and fantasies in order to be able to cope with life easier'. Even more, to bring knowledge of such truths into the practice of artistic engagement is necessary in order to understand character's emotional, psychological and situational aspects as described in the work (Carroll 2002: 4). But if we already know the truth of thematic statements, then ultimately artworks cannot be said to promote propositional knowledge.

Second, this approach also fails to meet the no-justification objection. Propositional knowledge requires justified true beliefs; yet, works of art cannot provide such justification on the account of being fictional, make-believe creations (Stolnitz 1992, Lamarque and Olsen 1994, Graham 1995, Gibson 2003). While this objection allows the possibility of works of art conveying non-trivial truths about the real world, it denies their ability to provide justification for such truths, and propositional knowledge necessarily demands justification. Presumably, works of art fail to provide justification for the conveyed true beliefs by means of all three main sources: empirical evidence, argumentation and authorial testimony. First, given that works of art represent the fictional world of author's imagination, they cannot serve as a reliable source of empirical evidence. To borrow an example given by Jerome Stolnitz, while Dostoevsky's fictional novel *Crime and Punishment* offers a non-trivially true thematic statement that 'criminals desire to be caught and punished', the work itself cannot provide evidentiary support for the truthfulness of such claim (1992: 199–200). Artistic statements referring to the real world can only be supported by the real-world evidence. Hence, even though Dostoyevsky establishes a possible truth-claim about the real world, such claim can only find justification outside work in further scientific studies and empirical testing. This objection equally applies to both partly non-fictional and wholly non-fictional artworks. An example of the former is historical medical drama television show *The Knick* (2014), which incorporates several historical and medical facts, such as the accurate description of the first rhinoplasty procedure. The work can bring to light information that we did not know beforehand; yet, it cannot provide evidentiary support for such information for without consulting external historical resources we cannot be sure what is fictional and what not. An example of the wholly non-fictional work of art is Augusten Burroughs's memoir *Running with Scissors* (2002), which tells a true story of author's childhood life, after his emotionally unstable mother sends him to live with her psychiatrist Dr Finch. The author describes Dr Finch as a bizarre figure for he believes that God is communicating with him through his faeces and he develops a form of divination for trying to decipher these messages from God. While the work represents a true testimony of the author and can thus serve as a reliable source of information about Burrough's personal life, it cannot provide evidentiary justification for the general thematic statements that one can infer from it. For instance, one cannot take author's description of the psychiatrist as a reliable source of information that can be applicable to all psychiatrists.

Neither can artworks provide justification by means of argumentation given that their narratives typically lack cognitive tools such as reasoning, critical analysis and logical structure in the form of justified premised and conclusions. That is to say, the artwork itself does not support the truthfulness of its thematic statement by means of argumentative evidence (Lamarque and Olsen 1994: 332, Carroll 2002, Gibson 2003). Instead of serving the role of epistemic persuasion, artistic thematic statements function primarily to organize particular narrative details into a unified pattern for the purpose of providing overall meaning to the work. While Noel Carroll does consider some exceptional works of art that presumably provide argumentative analysis for their thematic claims (such as novels written by Ayn Rand), he nevertheless denies the idea that such work can be seen as argumentative works for the reason that they fail to articulate purported arguments by means of their distinctive narrative and formal features. As he writes:

> Obviously inasmuch as literary works are linguistic, they could contain full-blooded arguments, set forth step by step from premises to conclusions. But they are able to do this in virtue of their linguistic nature, not in virtue of their literary forms and structures. Just as a movie of a philosopher reciting an argument on screen would not be regarded as a cinematic argument – but merely an argument delivered by means of a motion picture – an argument stated prosaically in a novel, as baldly as it might be laid out in a logic textbook, would not count as a literary argument, that is, an argument made by means of literary forms, structures, and/ or devices.

(2018: 29)

Carroll's argument points towards another objection raised against propositional knowledge approach, namely, the textual-constraint objection as introduced by John Gibson. In short, the objection can be formulated in the following way: in order for artworks to have cognitive value, such as value of offering propositional knowledge, such knowledge must be part of the artwork's content. Factual truths must be bound up with the artwork's narrative details. For if knowledge is not given expression in the content of the work itself, then, as Gibson writes, 'we cannot claim to have learned that point from the work' (2009: 473). Yet, according to propositional knowledge approach, cognitive value of artworks lies in true thematic statements extracted from the fictional content. Recall that in order to avoid the paradox of fiction, propositional knowledge approach proposes a distinction between the artwork's subject level (work's fictional content) and its thematic level (thematic statements that refer to the real world). By proposing this move, however, such approach reduces work's cognitive value to a single thematic statement stripped of the work's narrative context. As Gibson states, propositional knowledge approach

> requires that we sever the stances, themes, and perspectives we find in a work from their literary context and treat them as free-floating propositions, asking what they might tell us about reality if we disregard their place and function in the text and instead treat them as isolated assertions about the way the world is.
> (Gibson 2003: 230)

If works of art are to be valued for their promotion of propositional knowledge, then that knowledge must be integrally connected to the content of the work. Otherwise, we cannot claim that the artwork qua artwork possesses cognitive value. The cognitive insight must be acquired through our engagement with the content of the work and its narrative details.[2]

The importance of this claim is furthered by Jenefer Robinson, who argues that our engagement with the narrative details is fundamental for learning from works of art: 'We cannot abstract the "message" of a great novel, because it is only through experiencing it that one can learn from it' (1995: 213). Emotional engagement with characters and situations as described in work of art is emotionally educational for it allows us to focus our attention on the relevant psychological and situational aspects of characters. This attention, she writes, ultimately reflects our own personal desires and interests without which we would not be able to respond affectively to characters, and which consequently allow us to formulate our own critical interpretation of the fictional world. Without emotionally affective responses to particular narrative details we would not be able to properly understand the content of the work and its significance for our own emotional lives. Robinson gives some examples of how emotional education occurs. She writes that we would not be able to notice Anna Karenina's situation and her vulnerability unless we affectively experience sympathy and compassion for her, or Macbeth's unethical character unless we experience repulsion towards him. Such emotional responses form the basis of our critical engagement with the artwork: 'Our emotional experience of the novel or play is itself a form of understanding, even if it is an inarticulate or relatively inarticulate understanding: if I laugh and cry, shiver, tense, and relax in all the appropriate places, then I can be said to have understood the story' (2005: 123). While Robinson does not advocate propositional knowledge approach, but rather emotional education, she nevertheless nicely illustrates the textual-constraint requirement for any form of aesthetic cognitivism and the important role that narrative details play in our cognitive experience of artworks.

Finally, given that propositional knowledge approach reduces cognitive value of artworks to a thematic statement devoid of work's fictional content, and therefore devoid of aesthetic-formal features that provide narrative structure, it also fails to meet the aesthetic-relevance objection. As pointed out previously, aesthetic cognitivism is the view arguing not only that works of art have cognitive value, but also that cognitive value is aesthetically relevant, i.e. works of art are aesthetically better because of the cognitive value they possess. But cognitive value can count as aesthetically relevant only when it is 'expressed by artistic means; it is the way, or mode, in which ethical and other insights are conveyed that makes them of relevance', whereby what counts as artistic means are 'factors such as the development of a persona, the way that insights are integrated into the particulars dealt with by works, and the way that we are affectively moved to feel the force of these insights' (Gaut 2007: 170). In other words, cognitive value that artworks possess determines their aesthetic value when it is necessitated by the aesthetic properties of the work – such as relevant formal and stylistic aspects, sensory elements, representational and expressive properties, which, when taken together, provide a heightened aesthetic enjoyment or appreciation of the work. But cognitive value of general thematic statements is separated from artwork's narrative and aesthetic aspects, and thus aesthetically irrelevant.

In an attempt to salvage propositional knowledge approach from these objections, aesthetic cognitivists have responded by offering several solutions. One solution argues that works of art offer not true thematic statements, but rather live hypotheses that the audience comes to confirm (or disconfirm) in the afterlife of the fictional work, the afterlife referring to the 'thinking about, remembering, reimagining what we have experienced in the gaps or after completing a good novel' (Kivy 1997: 133).[3] According to this account, it is the task of the audience, rather than the work itself, to provide evidentiary and argumentative support for the live hypotheses offered by the work (Carroll 2018: 27). The audience itself tests the hypotheses by means of their own experiences and testimonial evidence of others. While this approach can avoid the no-justification argument, it nevertheless fails to meet other anti-cognitivists objections.[4] In particular, it fails to accommodate the textual-constraint and aesthetic relevance objection, as pointed out by John Gibson:

> When we convert a theme into the form of a hypothesis (or proposition), we are already at one remove from the work; and when we begin scrutinizing this hypothesis for truth, we soon find ourselves at a second. And it seems clear that the labour put into forging the connection to truth is, again, performed almost entirely by the reader rather than the work, and thus we face the old problem.
> (2009: 472)

A different solution, given by Stacie Friend, argues that while works of art offer mere trivial truths, they can nevertheless enhance our comprehension and integration of these truths. Acquisition of knowledge consists not in mere 'accumulation of true beliefs, the storing of a list of facts in a little black box in the mind' but also in the 'integration of new information with old, organized so that it can be applied in other contexts' (2007: 40). Works of art employ distinct narrative devices to convey factual information, such as describing depicted events from the first-person perspective or offering concrete depiction of characters and their experiences. Since such devices generate rich mental images and facilitate more intimate and emotional engagements with the story, they can promote better comprehension of information and memorability. Finally, the third and most widely held solution offered by aesthetic cognitivism is to argue in favour of the idea that works of art convey non-propositional, rather than propositional (or factual) knowledge. Presumably, artwork's cognitive value is grounded not only in its ability to deliver factual truths, but also in its ability to promote empathic skills, improving our established conceptual vocabulary and in fostering the understanding of the significance, values and consequences that mere knowledge of something has in relation to human experiences. I will discuss each of these approaches respectively in what follows.

Experiential knowledge approach

According to this approach, works of art serve as a source of experiential or subjective knowledge (also known as knowledge by acquaintance), which refers to knowing what a certain experience is like, say, knowing what it is like to be jealous, homeless or to experience the horrors of war (Walsh 1969, Schick 1982, Wilson 1983, Currie 1995, Kieran 1996, Gaut 2006, Stroud 2008, Kajtar 2016). Experiential knowledge consists in knowing the qualitative character of certain conscious experiences or the what-it-is-like character of mental states, which cannot be exhausted by mere propositional knowledge.[5] For example, one can have propositional knowledge that 'love can quickly convert into hate', but if one has never actually experienced love and its complex manifestations, then one does not have experiential knowledge of what it feels like to experience love's alteration to hate. A first-personal acquaintance with a certain mental state is required to obtain experiential knowledge. While works of art cannot give us direct, actual experiences of mental states, they can nevertheless approximate such experiences. That is to say, they can offer 'a more complete notion of what it would be like to actually be having that experience' (Stroud 2008: 28). Presumably, works of art can promote such knowledge by means offering detailed, rich and vivid description of fictional characters and their situational aspects, which in turn can facilitate a certain type of imaginative activity in the audience, namely, the activity of mental simulation. Mental simulation refers to our ability to 'project ourselves in imagination into the situation of others' (Currie and Ravenscroft 2002: 51) and to mentalize or imagine what the other person is feeling, thinking, believing and desiring.

By means of mental simulation, works of art can promote the acquisition of two types of experiential knowledge. First, knowledge what it is like or would be like for me to be in another person's situation, say, knowledge what it would be like for me to suddenly suffer from anterograde amnesia and are unable to store new memories (as the main character did in the Christopher Nolan's film *Memento*, 2000). Knowing what it is like for me to be in another person's situation depends on a self-oriented perspective-taking (Coplan 2011: 9). One simply takes one's own perspective, together with the emotional and psychological traits one has, to mentalize the situation someone else is experiencing. While self-oriented perspective taking does not yield knowledge what it is like to be another person, it can nevertheless promote knowledge what-it-would-be-like for me to experience something, say, what it would be like for me to experience homelessness or jealousy. Such form of experiential knowledge can be highly beneficial for our own well-being as it allows us to imaginatively experience various situations we might never encounter in our lives, thereby giving us the opportunity to recognize and reconsider certain beliefs and values we momentarily hold. In this respect, self-oriented perspective taking can lead to the acquisition of self-knowledge and self-change.[6] Second, mental simulation can also give us the opportunity to obtain knowledge what it is like for another person to be in a certain situation, say, what it is like for the main protagonist in the film *Memento* to experience the loss of memory. This imaginative ability is called other-oriented perspective-taking or empathy: 'In other-oriented perspective-taking, when I successfully adopt the target's perspective, I imagine being the target undergoing the target's experiences rather than imagining being myself undergoing the target's experiences' (Coplan 2011: 13). Here we suspend our own personality in order to imagine that we are the other person, possessing their emotional and psychological traits, and assess the situation from their perspective. By imagining what it is like to be another person – a person that most likely has different feelings, thoughts and experiences – we can come to obtain greater understanding, acceptance and appreciation of a perspective of people that differ from us. Such understanding can consequently lessen our social and personal prejudices (Feshbach and Feshbach 2009), enable better prediction of other people's responses, reactions and behaviours, and thus significantly improve our communication and interpretation skills (Neill 2006, Batson 2009). Alex Neill nicely describes the cognitive effect of the other-oriented-perspective taking (or empathy) as following: 'In coming to see things as others see them and to feel as they do we gain a broader perspective on the world, an increased awareness and understanding of the possible modes of response to the world. In short, through responding empathetically to others we may come to see our world and our possibilities anew' (2006: 258).

Given the various socio-epistemic effects of the other-oriented perspective-taking, it is not surprising that proponents of the experiential knowledge approach focus on this form of knowledge as the main source of artwork's cognitive value. As the argument goes, works of art are particularly well suited to promote experiential knowledge what it is like to be another person for they typically offer detailed information about the character's emotional, psychological and situational aspects, this information being necessary for the activation of other-oriented perspective-taking (Goldie 2000: 195, Coplan 2011).

Yet, it is one thing to claim that works of art offer knowledge what it is like to be a particular fictional character and another thing to claim that knowledge thus gained also promotes our understanding of people in real life. That is to say, advocates of this approach must provide a legitimate defence of the claim that mental simulation of fictional character's emotional and situational aspects also leads to the acquisition of knowledge that the audience can apply to real-life situations and people. Presumably, such defence is found in the claim that fictional characters and situations are similar to people and situations in real life. What follows is the reconstruction of the proposed argument:

1. Having information about the character's situational, psychological and emotional aspects is necessary for the activation of other-oriented perspective-taking (or empathy).
2. Works of art typically provide us with great information about the psychological, emotional and situational aspects of fictional characters.
3. Hence, works of art promote the activation of other-oriented perspective-taking (or empathy) when engaging with fictional characters.
4. The activation of other-oriented perspective-taking (or empathy) leads to the acquisition of experiential knowledge what it is like to be a fictional character.
5. Fictional characters and their situational aspects are similar to people and situations in the real world.
6. Hence, knowing what it is like to be a fictional character leads to the acquisition of knowledge what it is like to be another person in real life.

The argument as it stands appears to be valid. In real life we often fail to mentalize other person's perspective for we do not have direct access to their mental processes. Works of art, on the other hand, can provide us with detailed information about the character's thoughts, motives, feelings and beliefs, which eases our ability to mentalize the depicted situation from their perspective and thereby to acquire knowledge what it is like to be a particular fictional character. It is easier to imagine things from another's point of view, if the mental processes of that person are explained. Moreover, given the affinity between fictional characters and real people, knowing what it is like to be a particular fictional character can help us understand people and their experiences in real life.

This approach can successfully meet the paradox of fiction. Even though works of art refer to imaginary or make-believe characters and situations, these characters nevertheless show emotional, psychological and situational affinity with real people. After all, without such perceived affinity, we would not be able to emotionally engage with them (Hakemulder 2000: 70–3). That is, perception of emotional, psychological and situational similarities (dissimilarities) with fictional characters facilitates (hinders) the activation of mental simulation. Furthermore, given that the knowledge obtained from works of art is not propositional, but rather experiential knowledge of what it is like to experience certain mental states, this approach can also avoid the cognitive triviality objection. As pointed out previously, possessing propositional knowledge that love can often convert into hate does not entail experiential knowledge of qualitative experiences, say, what love and its complex manifestations feel like. Works of art can thus promote new, non-trivial knowledge what it is like to experience certain mental states or state of affairs. Since such knowledge is not amenable to any sort of evidential or argumentative justification, but rather it depends solely on the conscious awareness of the person who directly experiences the what-it-is-like character of mental states, this approach can successfully meet the no-justification objection (Lamarque and Olsen 1994: 372, Stroud 2008: 22).[7] Finally, given that the artwork's ability to mentally simulate character's mental states is to a large extent necessitated by the aesthetic features of the work, such as formal and stylistic properties, as well as author's storytelling abilities of providing imaginative force to their characters, this suggests that cognitive value thus obtained is aesthetically relevant. The more vivid and detailed is work's description of fictional characters and their psychological and situational aspects, the easier it is to engage in mental simulation and consequently to obtain more acute experiential knowledge.

Nonetheless, while experiential knowledge approach can meet most of the objections given by anti-cognitivists, it is not fully successful as it raises many difficulties. Some of its difficulties have already been pointed out by John Gibson (2009), particularly important among which is his criticism of the approach's failure to meet the textual-constraint objection. As he writes: 'If the acquisition of truth depends on the reader applying aspects of a fictional world to the real one, then presumably that truth is not given expression in the work itself, and so the work cannot quite put it on offer' (2009: 471). Presumably, the problem is that experiential knowledge what it is like to be another person in the real world is not actually displayed within the artwork itself; rather, it is the audience that must exercise their own imaginative abilities of mentalizing what the fictional character is feeling, thinking, believing, desiring, and then apply this knowledge to the real world in order to perceive the human nature more clearly. But if it is the audience, rather than the content of the artwork itself that does the cognitive inquiry of knowing what it is like to be a real person, then it is not the artwork qua artwork that promotes experiential knowledge. As Gibson concludes elsewhere:

> There does seem to be a bit too much 'me' in all of this, and cognitivism is, again, about what goes on in artworks and not in the mind of the consumer about art [...] If what endows these simulated experiences with cognitive value is that I ask 'what would I think or feel if ... ', I am talking about my cognitive discovery, not an artwork's. No sensible person will deny that we can, and do, enlist artworks in our personal cognitive pursuits, and that artworks can have, if we just find a clever way of using them, instrumental cognitive value. But the simulation approach does not seem to take seriously that the question since Plato has been whether artworks themselves embody or articulate a kind of knowledge.
>
> (2008: 584–5)

Furthermore, many writers have criticized the premise 5 of the argument, namely, the thesis of the affinity between fictional characters and real people. The worry is that experiential knowledge acquired from works of art is closely tied to the specific and concrete nature of fictional situations, to characters that exhibit very unique psychological and emotional profiles, and thus that such knowledge lacks the epistemic force when applied to people and situations we encounter in real lives. For example, John Gibson argues that works of art fail to yield any substantive knowledge what it is like to be another person in real life since they typically portray extraordinary characters that differ greatly from people we meet in our daily lives: 'What kind of experiential knowledge do I "learn" from, say, Oedipus? – what it is like to have fate conspire against me such that I kill my father and sleep with my mother and in the process win and lose a kingdom? Surely this is not a kind of human experience' (2008: 583). Deborah Knight argues similarly for she claims that one's mental simulation of fictional characters is influenced by a whole range of formal and stylistic features of the work of art, which play no function in our understanding of real people:

> Whether something is actual or fictional will have a bearing on the sorts of explanations we produce ... Think of everything that becomes relevant if our target is a fictional character – things that have no conceivable role to play in our attempts to make sense of people. I am thinking, for example, of the genre or mode of the fiction; dominant themes, images, and motifs; the rhythm and/or pacing of the narrative; and in the case of film, mise-en-scè`ne, sound, music, framing and camera movement, the contribution of the personae of the actors to the realization of characters, and so on. We have no access to any understanding

of fictional characters in the absence of such features as these, because without these features, we have no access to fictional characters, period.

(2006: 275)

A different kind of criticism is given by Richard Gaskin, who rejects the idea that works of art can convey experiential knowledge what a certain experience is like. At best, he writes, they can give us knowledge what it is like to be reading or viewing about what a certain experience is like and this is a substantially different experience. As he states this objection in respect to literary art:

> the experience of sitting indoors reading could hardly be more unlike what it would be to experience the sequence of events that unfold in the novel" and hence "The activity of sitting at home and reading an exciting novel does not show you what it would be like to experience the events recounted in that novel; at most it shows you what it is like to sit at home reading an exciting novel about those events, which is a very different matter.

(2013: 121)

In addition to the textual-constraint objection and highly criticized affinity thesis, the experiential knowledge approach also faces a question as to what extent is our mental simulation of fictional characters representative of the other-perspective taking and if it is, how accurate such perspective taking can be. Namely, as pointed out by numerous research studies, it is one thing to activate other-oriented perspective, yet another to retain its accuracy, i.e. the ability to accurately infer other people's mental states during the process of mental simulation (Nickerson 1999, Epley and Caruso 2009, Coplan 2011). Mere activation of other-oriented perspective does not necessarily lead to the acquisition of knowledge what it is like to be another person. For example, Amy Coplan argues that following three conditions must be met in order to acquire experiential knowledge what it is like to be another person (empathic knowledge): (i) the affective state of an empathizer must match the affective state of the target person. For instance, if the target person experiences fear, then the empathizer must also experience the same affective state (though not necessarily in the same degree); (ii) the affective state of the empathizer must be the result of the other-oriented perspective-taking, that is, the empathizer must 'represent the other's situation from the other person's point of view and thus attempts to simulate the target's individual's experiences as though she were the target individual' (2011: 10). In other words, the empathizer must imagine that they are the other person, possessing the other person's emotional and psychological traits; and (iii) the empathizer must maintain self-other differentiation, that is, the empathizer must 'separate one's awareness of oneself and one's own experiences from one's representations of the other and the other's experiences' (2011: 16). The question is to what extent our engagement with fictional characters meets these three conditions. I will examine this question by analysing our emotional engagement with the plot and fictional characters in films, paintings and literature, respectively.

Noel Carroll has written most extensively on the topic of character engagement in cinematography and he concludes that our emotional engagement with fictional characters in films is not one of empathy, but rather of sympathy. This is evident from the observation that typically our emotional responses do not match the responses of the fictional characters themselves, thereby failing to satisfy the first condition required for obtaining knowledge what it is like to be another person. As he writes:

> When the heroine is splashing about with abandon as, unbeknownst to her, a killer shark is zooming in for the kill, we feel concern for her. But that is not what she is feeling. She's feeling delighted. [...] If we feel pity at Oedipus' recognition that he has killed his father and bedded his mother that is not what Oedipus is feeling. He is feeling guilt, remorse, and self-recrimination. And, needless to say, we are feeling none of these.
>
> (1990: 90–1)

The emotional asymmetry between the viewer and the fictional character often occurs in films that employ third-person perspective, which provides the viewer with more visual information than fictional characters have. In a third-person cinematic perspective the viewer takes the eyewitness perspective, i.e. looking from the outside in. One observes how the story unfolds *for* the character, rather than *with* the character, consequently experiencing feeling *for* the character, rather than *with* the character. The perspective one takes profoundly affects one's interpretation of fictional characters and their situational aspects. Given that engaging with the visual world from the third-person perspective provides the viewer with more information about the surrounding environment than the fictional characters themselves have, this information often intentionally used in the horror movie genre in order to provoke the suspense reaction, it follows that one's emotional experience will substantially differ from the experience of the fictional characters. Exceptions are films that employ first-person cinematic perspective, which represents the perspective of the character in the film (an example is the American horror film *The Blair Witch Project*, 1999). The viewer hereby observes the events through the eyes of the character, which creates more immersive experience for the viewer, as well as a more personal connection to characters and their situational aspects. Since the viewer's perspective is restricted to the perspective of the character, that is, they are both subjected to the same situational information, this enables their emotional matching. If the character responds, say, with fear by the danger lurking in the mist, so does the viewer. However, one might argue that while first-person cinematic perspective satisfies the emotional-matching condition, it is amenable to fail the requirement of other-oriented perspective-taking. As pointed out previously, to acquire knowledge what it is like to be another person, one's emotional response must not merely match the fictional character's response, but also that such emotional matching must result from the other-oriented perspective-taking, that is, one must imagine the perspective of the character and mentally simulate their experiences. Yet, films that employ first-person cinematic perspective necessitate emotional matching not because the viewer imagines the situation from the character's perspective, but rather because they are subjected to the same situational information. The viewer experiences the same emotion as the fictional character, say, the emotion of fear, because they perceive the same situational information, namely, the danger lurking in the mist.

Another example of emotional matching, without being empathy, is emotional contagion. In short, emotional contagion is an automatic and unconscious imitation or mimicking of other's emotional expressions. It is typically explained as a 'tendency to automatically mimic and synchronize facial expressions, vocalizations, postures, and movements with those of another person's and, consequently, to converge emotionally' (Hatfield, Cacioppo and Rapson 1994: 5). Because emotional contagion depends on a 'direct sensory stimulation and subsequent physiological responses to that stimulation' (Coplan 2006: 31), it does not require any information about another person's state of mind (Goldie 2000: 190–1, Coplan 2006). An example of emotional contagion is a scene in Alfred Hitchcock's film *Psycho* (1960), where Norman Bates (Anthony Perkins) puts Marion's (Janet Leigh) dead body in her car and sinks it in a swamp. As he impatiently waits for the car to sink into the lake, one perceives his anxious facial expression and a subtle look of relief as the car finally sinks, which triggers the activation of the facial configuration associated with the observed emotion (mimicry). This in turn induces in the viewer the physiological activations and the subjective experience of the very same emotion. Emotional contagion appears to be a dominant emotional response in cinematic artworks that use specific cinematographic techniques such as close-up of characters facial expressions and slow shots of bodily movements in order to minimize the narrative elements and highlight formal and aesthetics properties of the film. Minimal narration and lack of substantial information regarding the character's situational and psychological aspects hinder viewer's ability to activate other-oriented perspective.

The case appears to be similar in our emotional engagement with characters and situations depicted in paintings as they do not provide sufficient insight into the portrayed character's thoughts and feelings. Typically, our emotional response in respect to pictorial art is either emotional contagion or self-oriented perspective taking. An example of the former is Edvard Munch's painting *The Scream* (1893), which immediately evokes the feelings of anxiety and agony in a viewer. These feelings are not the result of centrally imagining the mental state of the represented figure; rather, one simply catches the emotion and the mood of the painting through the depiction of a screaming face and the blood reddish colour of the sky. On the other hand, self-oriented perspective taking is our 'default mode of mentalizing' when trying to understand other people's mental states, but without possessing sufficient information about these states (Coplan 2011: 10). In self-oriented perspective taking we imagine ourselves, with our own psychological profile, being in another person's situation. We often take such perspective automatically when we are confronted with works of art that have a story to tell, yet do not offer any substantive information about the mental states of characters. As Peter Goldie writes: 'Both characterization and narrative are independently necessary for empathy: without the former there is no possibility of centrally imagining another, and without the latter, there is no narrative to experience' (2000: 198). While pictorial art may have some story to tell, it lacks sufficient characterization. As a result, we have difficulties activating other-oriented perspective (i.e. imagining what the character must be thinking, feeling or believing) and thus automatically shift to adopting self-oriented perspective; we project some of our own thoughts, feelings and beliefs into the character. But to imagine the character's state of mind based on our own characteristics, beliefs and experiences can lead to serious misunderstandings of the characters and their situational aspects. Consider, for example, Vincent van Gogh's painting *Sorrowing Old Man: At Eternity's Gate* (1890), which depicts an old man, sitting on the chair with his elbows on his knees, burying his face in his hands. Without being acquainted with the title of the work, one might easily imagine that instead of feeling sorrow the old man is resting his eyes or he might be secretly laughing. Even if one knows that the painting represents the sorrow of an old age, one might still find it difficult to mentalize the state of mind of the depicted man, if not given additional information. One can always ask – how does it feel like to be burdened by the heaviness of an old age? If the question is asked by a person of young age, rich with goals, desires and aspirations, then the answer must be that it feels desperate and miserable. Yet, a person with maturity, whose values, desires and priorities are changed due to ageing, might not share the same emotional reaction. This example shows how lack of sufficient character information can quickly result in incorrect prediction and misunderstandings of other person's experiences. In addition, it demonstrates that the more dissimilar one's state of mind from the target person is, the more difficult it is to activate other-oriented perspective-taking.

However, the case is different with literary art, which tends to provide a strong narrative context with detailed description of characters' mental state and their relationship with others (Kuiken et al. 2004). This information substantially eases reader's transportation into the characters' state of mind and facilitates the activation of other-oriented perspective. Furthermore, given that literary art does not depend on a direct sensory perception, it precludes the danger of reader's emotional matching being the result of emotional contagion. It is not surprising, accordingly, that most studies of experiential knowledge in respect to art have been done in connection with literature.

Nonetheless, as pointed out by numerous research studies, there is a substantial difference between the activation of other-oriented perspective-taking and its accuracy. These studies show that activation of other-oriented perspective-taking is facilitated or hindered when engaging with people that we perceive emotionally and psychologically similar or dissimilar respectively (Hakemulder 2000: 70–3, Green 2004, Green and Donahue 2009). Furthermore, that engagement with people that we perceive characteristically similar tends to lead to inaccuracy of other-oriented perspective-taking. The reason for this phenomenon is that once we assume greater psychological and emotional similarity with the target person, which is the element that eases other-oriented perspective-taking, we tend to automatically shift to adopting egocentric or self-oriented perspective, the latter eventually overriding the initial other-oriented perspective (Ames 2004, Epley and Caruso 2009, Nickerson, Butler and Carlin 2009). It has been shown that first information highly influences our further perception and evaluation of others. In other words, if we initially perceive the other person as characteristically similar to us, then we will continue to project our own personal data to the target person without acknowledging that the other person might not completely share our perspective: 'This primacy may have such a profound effect on people's perceptions of an event that it may not even occur to them that others' perceptions may differ from their own and may therefore be in need of adjustment' (Epley and Caruso 2009: 300). But this implies that other-oriented perspective-taking is to a certain extent always contaminated with self-oriented perspective and with the projection of our own emotional and psychological characteristics into the target person. This consequently leads to misattributions, prediction errors and inaccurate perception of others, that is, to empathic inaccuracy.

Accordingly, even though works of art, especially of literary kind, do allow us to engage in other-oriented perspective-taking, there is nevertheless no guarantee that such perspective will lead to emphatic accuracy. When engaging with fictional characters we can never really be sure that our own perspective is left behind:

> When one attempts to imagine what it is like to be a specific other person, what one is really doing is imagining what it would be like to be oneself – how one would feel or behave – in the other person's situation … One can never be certain that one's own imagined experience in the imagined situation would be, in fact, the same at that of another person who is actually in that situation. The assumption of a close correspondence seems essential to empathy, but it is also important to recognize that the assumption could be wrong in many specific instances.
>
> (Nickerson, Butler and Carlin 2009: 52)

Furthermore, the perception of similarity with fictional characters is vulnerable to the failure of preserving clear self-other differentiation, which can lead to the occurrence of personal distress. Personal distress is the result of over-projection of our own thoughts and emotions into the fictional character, that is, we come to experience the character and the depicted situation as our own. This is the situation in which we focus our attention on our own experience, rather than on the experience of the fictional character, thereby disrupting a detached other-oriented perspective and failing to accurately represent the character's experience. As Amy Coplan points out:

> Imagining what it would be like for me to be in the awful situation you're experiencing makes it harder for me to modulate my emotions. I lose track of the fact that the experiences are actually yours and not mine and end up feeling so upset that I become completely focused on my own pain and what I can do to alleviate it. My emotional responses to imagined scenarios involving me as me lead to greater emotional arousal in general.
>
> (2011: 12)

In conclusion, artistic acquisition of experiential knowledge what it is like to be another person depends primarily on the activation of other-oriented perspective-taking, which is facilitated by the assumption of similarity between the audience and the fictional character. Yet, the assumption of similarity often leads to overriding other-oriented perspective-taking with the self-oriented, egocentric perspective, resulting thereby empathic inaccuracy. On the other hand, the assumption of dissimilarity with fictional characters hinders the activation of other-oriented perspective-taking. It is difficult to mentally simulate the experience of a person that is substantially different from us, undergoing a situation too remote from our own personal experiences. The acquisition of experiential knowledge what it is like to be a fictional character is often precluded either by the nature of a particular art form (such as forms of art that lack narrative details and substantive characterization) or by audience's psychology. Other-oriented perspective is a mentally challenging activity, requiring effortful thinking, mental flexibility and emotional self-regulation. It is influenced by the person's current mental and emotional state, their attentive abilities and their motivation to fully understand other person's perspective. But if works of art cannot provide, or provide with extreme difficulty, the accurate experiential knowledge what it is like to be a fictional character, then, a fortiori, they cannot provide, or provide with extreme difficulty experiential knowledge what it is like to be another person in real life.

Conceptual or philosophical knowledge approach

Proponents of this approach argue that there is a similarity between artworks and traditional philosophical texts in that they are both concerned with the exploration of perennial issues on the nature of human existence (such as the issue of truth, personal identity or free will). Moreover, works of art resemble philosophy not only by exhibiting philosophically significant content, but also by employing standard philosophical methods of inquiry, such as raising philosophical questions, providing counterexamples to refute philosophical positions or set up philosophical hypothesis (Russell 2006). Presumably, as the prevailing view argues, works of art can be seen as a recourse to philosophical inquiry by means of resembling a form of an argument, namely, a thought experiment that we use in science and particularly in philosophy (Davenport 1983, Carroll 2002, John 2005, Elgin 2007, Swirski 2007, Wartenberg 2007, Dadlez 2009, Green 2017).

As the argument goes, works of art are akin to philosophical thought experiments in that they are imaginative, fictional and narrative in nature. For example, Plato's thought experiment *The Ring of Gyges* asks us to imagine the possibility of having a magic ring that makes one invisible and how this imaginative situation challenges our established beliefs, attitudes and commitments we have in the matters of justice and morality. But works of art can function similarly. For instance, James Whale's science fiction film *The Invisible Man* (1933) can be seen as an imaginative thought experiment questioning our moral beliefs. The movie tells a story of a scientist Jack Griffin (Claude Rains) who discovers an invisibility formula. While once a rational human being, after he turns himself invisible, he becomes depraved maniac and murderer. The narrative of the film can be used in a broader philosophical discussion to reflect on the role of social restriction in the conception of morality and to challenge our own moral attitudes. If philosophical thought experiments function as a source of conceptual or philosophical knowledge in the sense that they serve to clarify and refine our established conceptual schemes, then so can works of art. Accordingly, as Noel Carroll concludes, works of art can 'prompt us to contemplate, possibly clarify, and even reconfigure our conceptual commitments, thereby rendering our concepts newly meaningful' (2002: 7). Moreover, given that in contrast to philosophical thought experiments, works of art offer imaginatively richer, more elaborated, detailed and affective background information, they can thereby offer more comprehensive, careful and specific connections between various concepts (Carroll 2002: 19, Elgin 2007).

This approach can successfully avoid the paradox of fiction, as well as the cognitive triviality and no-justification objections. Even though a great amount of artworks are construed from fictional statements, the concepts and words that constitute these statements nevertheless have worldly reference (Carroll 2002). That is, even in fictional situations we employ our standard conceptual vocabulary that we use in our ordinary experience of the world. Furthermore, given that artworks presumably function as thought experiments and hence as a form of an argument they meet the no-justification objection. They can stimulate some sort of reflective process that is required for obtaining conceptual or philosophical knowledge and thus lead to the refinement or modification of our conceptual vocabulary. Since works of art considered as thought experiments depend on the conceptual vocabulary we already possess, yet, they contribute to its refinement and modification, their cognitive value is not trivial (Carroll 2002: 8–9).

Nonetheless, this approach is not fully satisfactory. Many critics attack the idea that works of art contain a structure similar enough to philosophical or scientific thought experiments so that similar cognitive effects can be derived from them (Wilkes 1988, Sorensen 1998, Egan 2016, Currie 2020). For example, Kathleen Wilkes claims: 'The author of fantasy sets out for us a new framework within which the events taking place are (given a dollop of suspended belief) intelligible. He is not setting out to enable us to draw conclusions about our theories and our concepts' (1988: 45–6). David Egan's criticism is even more elaborate as he points out various disanalogies between works of art and thought experiment, such as artwork's inability to articulate the connection between the particular imaginative narrative and its role in a broader philosophical discussion: 'A fictional narrative only becomes a thought experiment to the extent that it is deployed in an argument [...] it is only in their contribution to this larger argumentative structure that they get their distinctive cognitive payoff' (2016: 142). But how exactly do artworks fail to contribute to the larger argumentative structure? To answer this question, I shall consider in more details the distinction between philosophical thought experiments and works of art.

According to Tamar Szabo Gendler (2000: 21), scientific and philosophical thought experiments are distinctive in virtue of having the following tripartite structure: (i) description of an imaginary world, that is, a world which is similar in relevant aspects to the real world and in which an imaginary scenario occurs; (ii) an argument is offered in an attempt to give a correct evaluation of the imaginary scenario, that is, what would we say if that imaginary scenario occurs; and (iii) the evaluation of the imaginary scenario is taken to reveal a larger lesson. For example, the imaginary scenario in Plato's *The Ring of Gyges* thought experiment is having a magic ring that makes one invisible, whereas the evaluation of the scenario is that both just and the unjust man would act the same way, namely commit violent crimes have they each been given the magic ring. The larger lesson is revealed by pointing out that people are just not willingly, that is, because they believe that justice is intrinsically good, but rather because of necessity. People are just only because of their fear of being caught and punished. As a result of the larger lesson, our conception of justice is substantially improved or modified.

For artworks to serve as a source of conceptual knowledge, their narrative structure must be sufficiently similar to the tripartite structure of philosophical thought experiments. Consider again the James Whale's film *The Invisible Man*. The film offers the imaginary scenario, this being the discovery of the invisibility formula, as well as its evaluation, which consists in the narrative demonstration that once a rational human being, the scientist after he turns himself invisible, becomes a violent criminal. However, the film lacks the articulation of the larger philosophical lesson, namely, how the imaginary scenario relates to the question of morality and justice. But if the work fails to explicitly articulate the implications and epistemic challenges of the imaginary scenario, this articulation being the 'demonstrative force' (Gendler 2010: 32) of any thought experiment, then it necessarily fails to deliver desirable cognitive effects. According to Murray Smith (2006), works of art fail to function as full-blown thought experiments, even though they entertain philosophically relevant themes, because they serve a substantially different purpose. While philosophical thought experiments have an epistemic purpose, which fits their typical structure, such as notable simplicity of the narrative and explicit specification of the imaginative implications in order to maximize the epistemic effect, artworks, on the other hand, serve aesthetic and entertaining purposes and accordingly employ organizational structure that serves to engage audience, such as detailed narration, stylistic components, structural complexity, indeterminacy and emotional engagement (Nussbaum 1990: 47, Currie 2010: 143). Moreover, artwork's stylistic and aesthetic features tend to generate multitude of interpretative possibilities, which in turn can hinder rather than facilitate abstract reflections typically generated by thought experiments (Lamarque 2014: 78–9, Egan 2016, Currie 2020). As David Egan writes:

> the features that make for a good thought experiment often make for a bad story and vice versa. Where we might praise a thought experiment for clearly schematizing the abstract concepts it aims to treat, we might criticize a work of literature for being too schematic. Where we might praise a work of literature for its subtlety and nuance, we might criticize a thought experiment for being muddy and lacking precision. We look for singularity of focus and clear analogies between the concrete and the abstract in thought experiments, but in literary fictions we want a broader perspective and one not so readily reduced to a set of abstract relations. The features of a fictional narrative that make it laudable as a work of literature tend to get in the way if we want to treat it as a thought experiment. Some forms of argumentative criticism can have a place in literary criticism, but not the sort of argumentative criticism we apply to thought experiments.
>
> (2016: 147)

Furthermore, if works of art fail to present the philosophically relevant narrative content in a philosophically relevant way, namely, in a way to explicitly articulate conceptually relevant implications of the imaginary scenario, then they also fail to satisfy the textual-constraint argument. In other words, what is at issue here is the distinction between the artwork itself containing a philosophical argument that reveals the larger lesson and the audience using the imaginary narrative to interpret it in a philosophically interesting way. As John Gibson presents this problem in respect to literary art:

> When we examine literature, however, we find plot occurrences rather than premises, dramatic events rather than supporting evidence, aesthetic feats rather than philosophical analysis ... We can, of course, use a literary test in the pursuit of knowledge. If we allow ourselves to blow argumentation into a literary work, we will find that it offers endless ways of coming to know reality ... While manoeuvres such as this may aid in our pursuit of worldly truth, should we take this step we lose the literary work.
>
> (2003: 229–30)

According to Gibson, works of art can promote conceptual or philosophical knowledge, but not as works of art. That is to say, works can offer conceptual knowledge only by means of audience's employment of their own reflective capacities to bring to light philosophically relevant conclusions. It is the audience rather than work's own cognitive content that does the cognitive inquiry. This is evident from observation that not every viewer acquires philosophically relevant knowledge when engaging with, say, James Whale's film *The Invisible Man*. The film is seen as a philosophical thought experiment and serves as a source of conceptual knowledge only by an audience with philosophical interests, whereas for most of us it merely offers an entertaining story, rather than a philosophical articulation of its theme.[8] This means that work's cognitive value is used instrumentally to promote conceptual knowledge, but this is not appreciating the work qua artwork. As Lester H. Hunt points out: 'Ultimately, though, motion pictures may resemble the humble laboratory rat in that its cognitive value may be found in what we do with it, rather than in what it tells us' (2006: 404). Furthermore, since conceptual knowledge is not part of the narrative content itself, it cannot be said to be determined by its aesthetic-formal properties; thus, artwork's cognitive value remains to be aesthetically irrelevant.

John Gibson and axiological understanding

A more recent and promising account of aesthetic cognitivism is given by John Gibson. According to his view, cognitive value of art lies in revealing certain aspects of the world, which cannot be grasped by means of epistemic terms such as truth or knowledge, but rather by means of axiological understanding, that is, recognizing significance, value and consequences that mere knowledge of something has in relation to human experiences and the world. His approach depends on the distinction he makes between criterial understanding (or mere knowledge) and axiological understanding. The former refers to one's correct identification of the object and its application to the real world, whereas the latter is 'a matter of value-of how something counts as an object of concern, as a site of significance, of how and why it matters' (Gibson 2009: 480). Axiological understanding represents an act of acknowledgement of mere knowledge: 'An act of acknowledgment is a way of giving life to what it is that we know, of bringing it into the public world ... It places us in the world as agents who are responsible to the range of values and experiences that are the mark of human reality' (Gibson 2009: 481).

Gibson illustrates the distinction between criterial and axiological understanding with an example of a person who witnesses an accident, but instead of calling the ambulance or rush to help injured people, they simply walk away. While this person is in possession of criterial understanding as they know that the accident occurred, that wounded people are in pain and need medical assistance; they nevertheless fail to recognize the consequences, concerns and significance of someone being in pain and suffering. By lacking such axiological understanding, they fail to mobilize themselves and intervene to relieve one's pain. According to Gibson, the person's failure to understand what it means that people are in pain is not a mere moral failure, but rather a cognitive failure; they fail to recognize the 'dramatic structure' of suffering and its significance when embodied in a particular human situation. It is only when we possess axiological understanding that can we establish ourselves not as mere passive knowers, but as doers, that is, as agents and active participants in transfiguring the human world.

In order to acquire axiological understanding, Gibson writes, we need to see how our criterial understanding or mere knowledge of the world functions when placed in a particular case of human activity, and for this 'a picture, a vision, of human activity is necessary, not the elaboration of a concept or principle' (Gibson 2009). Works of art offer such a vision as they provide contextualization of our conceptual and propositional knowledge in a particular imaginative narrative; hence, they are especially well-suited to promote axiological understanding: 'The vision of life we find in literary narratives show us human practices and circumstances not from an abstracted, external perspective, but from the "inside of life", in its full dramatic form' (Gibson 2009: 482). Works of art give us the opportunity to observe how our mere knowledge, say, knowledge what suffering, jealousy or racism is, functions in a particular human situation, and how it is embodied in one's psychology, living experiences and behaviour patterns. Consequently, they allow us to recognize the significance, consequences and values that mere knowledge plays in the human world.

Gibson's account offers the most promising defence of aesthetic cognitivism as it successfully avoids all four objections given by anti-cognitivists. Rather than offering new truths about the world, artwork's cognitive function consists in connecting previously attained truths with emotional, social and moral dimensions, thereby promoting axiological understanding. This is the kind of cognitive value that is not trivial, nor can it be justified by means of rational tools such as evidence and argumentation; rather, a vision of particular human affairs is required. Moreover, given that the vision of a dramatic structure is carried forward by artwork's internal narrative properties (its fictional content), this approach meets the textual-constraint objection. Finally, since the relevant cognitive insight is closely tied to artwork's narrative dimension, which reflects author's choice of aesthetic-formal properties, work's cognitive success is bound up with its aesthetic value.

However, Gibson's account is not without its challenges. In particular, his approach cannot accommodate cognitivism in respect to works of art with minimal or no narration, which can be cognitively significant and express complex ideas and concepts in virtue of form alone. If acquisition of axiological understanding requires, as Gibson writes, 'precisely what literature is in a position to give it: narrative, a story of human activity' (2003: 235), then it follows that only narrative forms of art (that is, forms of art that offer detailed and vivid description of characters and their situational aspects) can possess cognitive value. But this makes Gibson's account too excluding as it fails to ascribe cognitive value to non-representational works of art. For example, Francis Bacon's work *Three Studies for a Crucifixion* (1944) uses minimal visual narration in its portrayal of distorted mutations of the human form, through which he nonetheless reveals the truth of the corruption of the human spirit. Even Jackson Pollock's abstract paintings, such as *Number 1* (1950), which does not represent anything, still manage to reveal something important about the human reality, namely, that it is often random, chaotic and unpredictable. Furthermore, one could also make an argument that Gibson's account fails to explain sufficiently the cognitive value of symbolic and conceptual art, which is rich in rhetoric and metaphorical meanings, yet lacks a particular story, that is, it does not offer a concrete vision of human life and practices. Consider, for example, Merret Oppenheim's conceptual sculpture *My Nurse* (1936), which is made of a pair of white, long-heeled shoes, tied together, topped with paper ruffles used to decorate a roasted chicken, and presented on a silver plate. The meaning of the work can be revealed only by means of interpreting images as a kind of metaphorical or symbolic representations. For instance, the pair of shoes indirectly representing the women's body (the shoes are composed in a way that resembles a woman lying on her back, with legs spread apart); moreover, since it is the image of the high-heel shoes, this in addition brings to mind the idea of a female body as a fetishized object of erotic desire, whereas the image of the silver plate evokes the idea of the consumption of a female body. The combination of these metaphorical representations generates complex mental imagining of the idea of the objectification of women. Work's representation of this idea is cognitively significant and valuable, even though it is not presented through a concrete vision of human affairs.

There is yet another problem with Gibson's account, referring to his unconvincing distinction between criterial and axiological understanding. According to his position, works of art have a unique ability to facilitate axiological understanding because they 'take what is dull, wooden, or tenuous in our understanding of how our words and our concepts unite us with our world and inject it with this essential vitality of understanding' (2003: 236). Gibson seems to assume that our acquisition of criterial understanding is cut of the world and that consequently our knowledge of ideas and concepts functions as empty notions that must be accompanied by axiological understanding in order to give 'shape, form, and structure to the range of values, concerns, and experiences that define human reality' (2009: 482). Presumably, axiological understanding as attained through works of art reveals the 'gap between mind and reality' and 'shows us that the concept of knowledge alone does not express understanding as it reaches all the way into the world' (2003: 236). But is our acquisition of criterial understanding really free of the living world? Is it not the case that our identification of objects as particular kinds of things proceeds precisely by means of grasping their particular shapes? Consider again the play *Who Is Afraid of Virginia Wolf* (1962), which offers a truthful thematic statement that 'love can quickly convert into hate'. We come to recognize a particular behaviour between George and Martha as an instantiation of this thematic statement precisely by engaging and witnessing a particular form of human experience – how their relationship is characterized by their constant criticizing, ridiculing, physically abusing and humiliating each other, but which they do not seem to take it seriously. This indicates that our acquisition of criterial understanding is to a certain extent always permeated with some form of axiological understanding. As inhabitants of the social and cultural human world we already have our mere knowledge contextualized – our knowledge is already in the world, connected with its social, emotional and psychological dimensions. But if our mere knowledge of the world is already endowed with axiological understanding to a certain extent, then art cannot deliver a new kind of understanding, as Gibson presumably claims. At best, art can deepen the kind of knowledge of the world we already have, for example, through offering a more detailed vision of that world that we seldom have access to in our ordinary lives.[9]

Towards a positive account of aesthetic cognitivism: Objectual understanding

The foregoing discussion showed that none of the approaches given by aesthetic cognitivists are fully successful. With the exception of Gibson's account, none of them presented a form of aesthetic cognitivism that can meet all four objections given by anti-cognitivists. Gibson's account is highly valuable as it shows that our cognitive engagement with the world is much more complex and sophisticated to fit the traditional epistemic notions of knowledge and truth. It consists not only in correct identification of the features of the world and obtaining factual truths, but also in other epistemically valuable goals, such as making sense of worldly truths, acknowledging their significance for the human experiences and establishing deeper connection between the content of these truths and the real world. However, Gibson's account is problematic in that it depends on the distinction between criterial and axiological understanding that is not genuinely attainable. If criterial understanding is never devoid of particular human reality, then what is unique cognitive contribution of axiological understanding? Gibson writes that it consists in 'bringing our world to view' (2003: 236), presenting our concepts to us 'as concrete forms of human engagement' and as 'very precisely shaped human situations' (2009: 482). Yet, these are all highly slippery cognitive terms that seldom count as genuine cognitive achievements. As Peter Lamarque writes, if we consider practices such as 'exploring aspects of experience', 'providing visual images', 'broadening horizons' or 'exploring and elaborating human ideals' as genuine cognitive practices, then everyone must be a cognitivist (2006: 129). Accordingly, what is missing from Gibson's account is an elaboration of the epistemic notion of understanding such that it can be placed within the scientific context. Science is the paradigmatic form of human knowledge and if the kind of understanding that art presumably gives rise to is to count as a genuine cognitive achievement, then it must be a form of cognition well secured within a scientific practice as well. Such notion of understanding, as substantially different from the epistemic concept of knowledge, has indeed been developed in recent years by a number of leading epistemologists such as Jonathan Kvanvig, Linda Zagzebski, Catherine Elgin and Christoph Baumberger. According to their proposal, cognitive progress consists not only in the acquisition of factual truths (i.e. justified true beliefs), but more importantly in the understanding of already acquired true beliefs, whereby they define such understanding as 'an internal grasping or appreciation of how the various elements in a body of information are related to each other in terms of explanatory, logical, probabilistic, and other kinds of relations' (Kvanvig 2003: 192).

Presumably, understanding involves not merely knowing that or why something is the case, but rather comprehension of the internal relations between different elements of the phenomenon. Such a form of an understanding is often referred to as an objectual understanding, which refers to an understanding of the integrated and comprehensive body of information or set of beliefs and propositions. Objectual understanding is often compared to explanatory understanding (or understanding why something is the case by means of offering an explanation) in that they both involve grasping explanatory relations in the body of information.[10] However, objectual understanding is distinctive and significantly broader in that (i) it is an understanding of a particular subject matter; one can understand a scientific theory, a person, an artwork, a picture, a piece of music, an animal or an abstract concept (Zagzebski 2001, Kvanvig 2003, Baumberger, Beisbart and Brun 2017, Grimm 2017), and (ii) it involves not only grasping causal relationships (as explanatory understanding), but also understanding the implications, effects, consequences and significance of the phenomena. For example, to have (objectual) understanding of another person's angriness involves not only comprehending the underlying causal mechanisms of their anger, that is, how different introspective, emotional and affective aspects that constitute one's particular anger are internally related to one another, how they interact with each other in terms of cause and effect, part and whole, but also acknowledging the implications, consequences and awareness of their significance. As Christoph Baumberger explains objectual understanding:

> Understanding a subject matter involves more than understanding why some fact about it obtains. Besides understanding why it occurs, understanding global warming involves, for instance, understanding what effects it will have, which relations it has to human activities and how far the temperature is likely to rise in future. As a result, objectual understanding involves grasping

more explanatory and other coherence-making relationships in a more comprehensive body of information.

(Baumberger 2011: 17)

One of the distinctive features of objectual understanding is that it is not aimed at truth or exactness, but rather comprehension. As Linda Zagzebski writes, objectual understanding involves 'seeing how the parts of that body of knowledge fit together' and it is achieved 'partly by simplifying what is understood, highlighting certain features and ignoring others' (2001: 244). Given that objectual understanding does not aim at delivering facts, it does not require that all of the beliefs comprising the body of information are true (Zagzebski 2001, Kvanvig 2003, Elgin 2009). Catherine Elgin nicely illustrates the non-facticity of objectual understanding by giving an example of a child who thinks that human beings descended from apes and argues that although the child's belief is not strictly speaking true, it does nevertheless exhibit some understanding of human evolution, such as understanding of the concept of evolution and the close relationship between humans and apes. As she writes, 'although there is a falsehood involved, it is a falsehood that enables her to connect, synthesize, and grasp a body of information that is grounded in the biological facts, and is supported (to an extent) by her available evidence' (2009: 329).

Since objectual understanding does not aim at truth, it does not require an appeal to external evidence for its justification; rather, objectual understanding is necessarily reflective and involves conscious transparency: 'Understanding has internalist conditions for success, whereas knowledge does not [...] It may be possible to know without knowing that one knows, but it is impossible to understand without understanding that one understands [...] understanding is a state in which I am directly aware of the object of my understanding, and conscious transparency is a criterion for understanding' (Zagzebski 2001: 246–7). We cannot understand a particular phenomenon without being aware of having achieved the state of understanding. One demonstrates the achievement of an understanding based on their own internal resources. Objectual understanding is accordingly considered as a property of a person and not of a proposition or a belief – it is a result of person's own mental ability of putting together pieces of information and comprehending the connections between them. To understand something, we must see or grasp the connections by ourselves and cannot rely on the testimony of experts: 'It is the internal seeing or appreciating of explanatory and other coherence-inducing relationships in a body of information that is crucial for understanding' (Kvanvig 2003: 198). Given that the grasp of the internal relationship between the body of information can be more or less comprehensive, objectual understanding, in contrast to knowledge, admits of degrees (Kvanvig 2003: 196, Elgin 2009: 324, Baumberger 2014). The difference in the degrees of the understanding depends on the degree of coherence and the amount of the relations taken into account. The more comprehensive one's grasp of the relationship between the body of information, the better is their understanding of the subject matter.

Furthermore, given that objectual understanding does not have as its object individual propositions, but rather comprehension of their internal relationship, it can be highly interpretative, that is, it can necessitate different perspectives depending on which elements in the body of information we make salient and which we ignore. Linda Zagzebski describes this feature of objectual understanding accordingly:

> Understanding does not always build on a base of knowledge. It may be achieved in more than one way about the same portion of reality. More than one alternative theory may give understanding of the same subject matter. This makes sense if we think of a theory as a representation of reality, where alternative representations can be better or worse, more or less accurate. But more than one may be equally good, equally accurate. This form of understanding does not presuppose knowledge or even true belief.
>
> (2001: 244)

Each representation of reality can weigh differently the significance of the elements or aspects of reality, revealing aspects that other representations obscure, yet both advancing a different understanding of the same phenomenon.

In sum, objectual understanding demands internal awareness of the relationship between different pieces of information, that is, how pieces of information belonging to a particular phenomenon are internally related with each other into a unified idea, rather than the facticity of a single belief. Given that objectual understanding refers to seeing and grasping the connections between different aspects of reality, it is right to say that it 'mirrors the world more profoundly than a mind which merely assents to propositions' (Grimm 2012: 110). It allows us to appreciate how different aspects of reality are related to one another and how they can be altered to achieve a certain desired outcome. Thus, objectual understanding not only 'deepens our cognitive grasp of that which is already known' (Zagzebski 2001: 244), but also enables us to make greater predictions about the world and thereby the ability to control it (Grimm 2017: 225).

One can notice that the epistemic notion of objectual understanding fits well with your cognitive experience of art, namely, that we often lack an adequate cognitive vocabulary to articulate the obtained cognitive insight. We feel that we have learned something important by engaging with works of art; yet, we have difficulties to communicate in words and propositions what exactly it is that we have learned. But, objectual understanding is precisely a form of cognitive state that can be expressed non-propositionally since it has as its object non-propositional structure of reality, namely, the internal seeing or awareness of the relationships between different aspects of the phenomena. As Linda Zagzebski writes: 'Understanding involves seeing how the parts of that body of knowledge fit together, where the fitting together is not itself propositional in form' (2001: 244). Even though narrative art has a propositional structure as it is composed of sequence of propositions, the relationship between these propositions (i.e. their formal structure) and consequently the internal grasping of such relationships cannot be grasped propositionally. But if we admit that cognitive appreciation of art lies in the internal seeing or grasping the structure of the object and thus in acquiring objectual understanding, then we can see in what way cognitive value of art can matter aesthetically. As pointed out previously, aesthetic appreciation is an appreciation of the structure of the work and how its formal properties, narrative content and thematic characterization cohere into a unified whole, thereby offering various perspectives on the subject matter. Accordingly, both cognitive and aesthetic appreciations have the same object of appreciation and invite the same mode of attention – the internal awareness of the formal relations between different elements of the phenomenon. In fact, as I will argue latter, artwork's aesthetic and cognitive value is intrinsically intertwined for it is precisely the comprehension of the various relationships between different aspects of the phenomena (objectual understanding) that gives rise to the aesthetically valuable feeling of pleasure.

The idea of understanding serving as a source of artwork's cognitive value has been in recent years acknowledged by various writers, such as Nelson Goodman (1978), Gordon Graham (2005), Jukka Mikkonen (2015), Catherine Elgin (2017), Christoph Baumberger (2013) and J. W. Phelan (2021). However, with an exception of Catherine Elgin, they introduce artistic understanding as a broader epistemic concept, involving not merely objectual understanding, but also other various epistemic values such as propositional, experiential, conceptual knowledge, as well as explanatory understanding. For example, Gordon Graham writes that artistic understanding is the kind of understanding that is employed in scientific endeavour as well and can be adequate or defective, more or less great, but not false. The difference between scientific and artistic understanding is that the former has as its object physical universe and the natural world, whereas the latter prioritizes concepts of human nature and human conditions. Presumably, works of art necessitate the understanding of human nature by means of providing us with imaginary representations that illuminate our everyday experience of the world: 'The imagination of the artist can transform our experience by enabling us to see, hear, touch, feel and think it more imaginatively, and thus enrich our understanding of it. It is in this sense that art is a source of understanding' (2005: 70). Similarly, Jukka Mikkonen argues that the central epistemic component of artworks is understanding, which consist in aspects such as developing 'readers' perception, provide them new perspectives on familiar things, help them acknowledge previously unnoticed relations between concepts, and offer them new categories for classifying objects' (2015: 274–5). Both Christoph Baumberger and J. W. Phelan claim that artworks contribute to the promotion of understanding in various ways, such as by providing new conceptual categories for objects, new hypotheses and beliefs that, when cohere with our established beliefs, offer new perspectival understanding of objects and state of affairs, apprehending different layers of meaning, grasping explanatory connections, as well as understanding as knowledge-how and knowledge what-it-is-like. None of these approaches, however, consider specifying the relationship between artworks and their ability to promote a distinctive type of understanding, namely, objectual understanding. Catherine Elgin is an exception as she defends the cognitive value of art in respect to objectual understanding that depends on the specifics of her own epistemological theory. According to her account, objectual understanding is constituted by a coherent set of epistemic commitments one holds towards pieces of information, whereas these pieces of information can take both propositional and non-propositional form; moreover, they are not necessarily factive – some of epistemic commitments that constitute objectual understanding may be false. Furthermore, epistemic commitments constitute objectual understanding as long as they are in a reflective equilibrium, which requires that one's 'epistemic commitments be mutually supportive and that they constitute an account that is at least as reasonable as any available alternative in the epistemic circumstances' (2017: 98–9). Works of art advance objectual understanding by 'improving on the commitments we currently hold, where improvement itself must be measured by current standards' (2017: 67). One of the ways that art can improve our epistemic commitments that constitute objectual understanding is by way of exemplification: 'When an item serves as a sample or example, it exemplifies: it functions as a symbol that makes reference to some of the properties, patterns, or relations it instantiates' (2017: 184). An item can exemplify a various amount of features it instantiates. By means of such exemplification, works of art render certain properties salient, thereby giving us the opportunity to appreciate features of the world, as well as their significance that we might have previously overlooked: 'Exemplification is not just a device for underscoring salient features or supplying emphasis. It often highlights and affords epistemic access to features that were previously disregarded, even to features that are semantically unmarked' (2017: 187). She nicely illustrates her account by means of dance exemplification. She writes that dance 'metaphorically exemplifies properties such as love and longing, weightlessness and ethereality' (2017: 208) by means of aesthetic properties of grace, beauty and delicacy, or it can exemplify emotional properties: 'One dancer droops, displaying a particular posture that metaphorically exemplifies grief. Another leaps, displaying a motion that metaphorically exemplifies joy' (2017: 215). Works of art can thereby help us recognize and appreciate certain features of the world that previously remained unnoticed and by integrating such features into the coherent set of epistemic commitment they can advance our understanding with respect to a certain phenomenon.

Elgin's account of the role that artistic exemplification plays in the promotion of objectual understanding is highly convincing and instructive. Artworks can, through exemplification, direct our attention to interesting properties of the world that we would otherwise overlook and thereby allow us to discover new ways of organizing relationships between different features of the object, and thus new ways of grasping and understanding the explanatory and inferential interrelationships between different features of the phenomenon. In defending aesthetic cognitivism in respect to art, I take a similar approach. I argue that cognitive value of art can and should be grasped within the epistemic framework of objectual understanding defined as the internal grasping or appreciations of the comprehensive structure of the object. Given that works of art deal essentially with highly abstract themes of philosophical, psychological, social and metaphysical kind, I argue that they are particularly well suited to promote objectual understanding of abstract phenomena in that they reveal how various introspective, emotional and affective information that appear to be central to the content of such phenomena are related to each other in terms of explanatory and inferential relations. I develop my proposal in light of Kant's theory of beautiful art as an expression of aesthetic ideas.

2

Kant and art as expression of aesthetic ideas

In the previous chapter I examined the main contemporary approaches given to defend aesthetic cognitivism and point out their inadequacies. My aim in this chapter is to propose a different account of aesthetic cognitivism – an account that takes into consideration Kant's theory of beautiful art as an expression of aesthetic ideas. On the account I propose, the cognitive value of artworks lies in promoting (objectual) understanding of abstract phenomena as they are determined by our own subjective experiences. I refer specifically to contemporary studies in cognitive science, which stress the importance of emotional, introspective and affective information for establishing the meaning of abstract concepts, such as concepts of truth, immortality, alienation, hopelessness and which lack a directly perceivable referent. In other words, understanding the semantic content of abstract concepts cannot be made in isolation of comprehending experiential information that significantly contributes to the meaning of these concepts. The gist of my argument is that works of art, conceived as expressions of aesthetic ideas, make salient the various causal and explanatory relationships between different experience-related features involved in our abstract concepts, thereby promoting their understanding. That is to say, aesthetic ideas make abstract concepts more cognitively accessible to us, by creating particular imaginative representations (that is, aesthetic attributes) that allow us to think about these concepts in a way linked to sensory experience. To develop my argument, I begin the first part of the chapter with a discussion of Kant's doctrine of aesthetic ideas and explain how aesthetic ideas can serve as indirect sensuous counterparts of abstract phenomena. Next, I discuss some of the issues pertaining to the apprehension of meaning and cognitive processing of abstract representations that have been of interest to cognitive science. Finally, I bring together the resources introduced and developed in the first part of the chapter and propose an account of the role of works of art in promoting understanding of abstract concepts.

Kant's theory of aesthetic ideas

In § 49 of the *Critique of the Power of Judgment* Kant puts forward his view of beautiful works of art functioning as expressions of aesthetic ideas. According to him, an aesthetic idea is a 'representation of the imagination that occasions much thinking, though without it being possible for any determinate thought, i.e., concept, to be adequate to it, which, consequently, no language fully attains or can make intelligible' (KU 5:314; 192). He also states in the same passage that an aesthetic idea is a 'counterpart (pendant) of an idea of reason, which is, conversely, a concept to which no intuition (representation of the imagination) can be adequate' (KU 5:314; 192). And later in §57 he adds that an aesthetic idea is 'an intuition (of the imagination) for which a concept can never be found adequate' (KU 5:342; 218).

Given Kant's characterization of (human) intuition as something that is sensible (A68/B93), singular and directly refers to the object (A320/B377), these passages suggest the view according to which aesthetic ideas are some kind of sensible representations of imagination, that is, images of some sort, and that these images are so rich and give rise to so much thinking that cannot be fully described by any determinate concepts.[1] Since aesthetic ideas lack adequate determinate concepts, they evade the possibility of cognition. That is to say, they cannot be cognized in an ordinary sense by connecting intuitions with determinate concepts. For this reason, they are also called ideas, even though they are strictly speaking sensible representations: 'Ideas in the most general meaning are representations related to an object in accordance with a certain (subjective or objective) principle, insofar as they can nevertheless never become a cognition of that object' (KU 5:342; 217).

In Kant's view, aesthetic ideas strive to exhibit concepts that go beyond the limits of our ordinary experience. He has two kinds of concepts in mind. On the one hand, there are concepts of reason (rational ideas), such as 'invisible beings, the kingdom of the blessed, the kingdom of hell, eternity, creation' (KU 5:314; 192). They are 'concept[s] to which no intuition (representation of imagination) can be adequate' (KU 5:314; 192). What is distinctive about them is that they can be thought, but not empirically encountered. For example, we can think of the idea of hell, but have no empirical intuition of it. That is, the content of the idea of hell cannot be experienced by the means of our senses. According to Kant, concepts that lack an adequate sensible correlate are called indeterminate concepts (in contrast to determinate concepts for which empirical intuition can be given) (KU 5:339; 215).

On the other hand, Kant writes that aesthetic ideas can also exhibit concepts 'of which there are examples in experience, e.g., death, envy, and all sorts of vices, as well as love, fame, etc., [...] with a completeness that goes beyond anything of which there is an example in nature' (KU 5:314; 192). These kinds of concepts, typically classified as abstract concepts, are dissimilar to rational ideas in that their concrete instances can be experienced.[2] For example, the abstract concept of truth is grounded in a particular situation such as a child confessing to their parents and the concept of (romantic) love is directly grounded in one's emotional and bodily experience (through behavioural signs like smiling, accelerated heartbeat, loss of appetite and so forth). On the other hand, they are also dissimilar to our ordinary (concrete) empirical concepts (such as dog, table, chair) in that they refer to non-material and non-concrete objects. For example, concepts such as love, envy and fame have no particular shape, size or colour, and one cannot see, touch or hear them. There is no single and concretely perceivable object that would correspond to such concepts. Their referents are mental and emotional states, personality traits, situations and events.

However, even though we can find examples of abstract concepts in ordinary experiences, their full meaning extends beyond such experience. This is because, as pointed out by contemporary studies in cognitive science, the content of abstract concepts involves to a large extent 'properties expressing subjective experiences' (Wiemer-Hastings and Xu 2005: 719), that is, experiential information, emotional aspects and other introspective properties, such as beliefs, memories, intentions, goals. For example, the content of the concept of alienation contains detailed features of the felt experience (what alienation feels like subjectively). But the felt experience, that is, the interplay of thoughts and feelings that lie at the background of our experience of alienation, cannot be directly exhibited in ordinary experience.[3] This is at least how I interpret Kant's decision to include abstract concepts in the category of concepts that aesthetic ideas strive to exhibit, in spite of his claim that aesthetic ideas strive to exhibit concepts that go beyond sensory experience. Thus, the implication seems to be that there is an additional, indeterminate material to these concepts for which no sensible intuition can be given and it is this material that aesthetic ideas strive to exhibit. That is to say, an aesthetic idea appears to function as an indirect sensible presentation of this additional and subjectively determined material that we associate with rational ideas and abstract concepts.

According to Kant, aesthetic ideas are generated by means of aesthetic attributes. He describes aesthetic attributes primarily in terms of what they do – namely, they 'express only the implications connected with it [a concept] and its affinity with others' (KU 5:315; 193). The function of aesthetic attributes is to bring to mind various connections or mental associations between different concepts and objects that, when taken together, give rise to a general idea (that is, rational ideas or abstract concepts). Kant gives an example of the image of an eagle with a lightning bolt in its talons, which functions as an aesthetic attribute of the rational idea of the king of heaven in the sense that it expresses certain associations and implications connected with this idea (in terms of representing power, strength, freedom, being above the material world) (KU 5:315; 193). These aesthetic attributes are partly determined by culturally established associations for we immediately connect the image of an eagle with the idea of heavenly king. As Aviv Reiter (2017) and Aaron Halper (2019) correctly point out, such culturally determined meanings of aesthetic attributes are required in order to ensure that the overall meaning of the artwork as intended by the artist is universally communicated.

Kant says very little, however, about what exactly aesthetic attributes are and how they come to occasion mental connections and associations. Fortunately, he offers a few remarks that can help us to formulate a plausible explanation. First, he writes that aesthetic attributes are 'supplementary representations of imagination' that 'go alongside the logical ones', which means that imagination produces them in addition to logical attributes. According to Kant, logical attributes 'constitute the presentation of a given concept itself' (KU 5:315; 193), that is, a schema of a determinate concept, and thus they refer to general representations that different objects of the same kind have in common and in virtue of which the determinate concept is applied.[4] Logical attributes play an essential role in our perceptual cognition. For example, recognizing a particular object as a dog involves applying logical attributes, such as an animal, four legs, fur and barking to the sensible manifold. Logical attributes simply refer to the set of marks common to different objects of the same kind and which constitute the content of a determinate concepts, such as concept of a dog. Given this definition, aesthetic attributes as supplementary or additional representations must refer to features of the object that go beyond these general features and which are not strictly speaking required for determinate cognition. Second, aesthetic attributes are 'attribute[s] of a representation of sense' (KU 5:316; 194) meaning that they must refer to features of an object with which we are directly perceptually acquainted. Third, they are product of 'the imagination, in its freedom from all guidance by rules' (KU 5:317; 195), which means that imagination produces them without being determined by the concept of the object.

Taking all these points together, the suggestion seems to be that aesthetic attributes refer to the distinctive aspects of a particular object, in contrast to those aspects of the object which are shared by all members of a class and in virtue of which the concept applies. Since these distinctive features of a particular object are not determined by the concepts of an object, they are a product of imagination in its free play. Let me explain this proposal in what follows.

According to Kant's epistemological theory, in order to have a perceptual image, conceptual harmony between imagination and understanding is necessary. We must perceive a certain combination of the sensible manifold under some empirical concepts. However, even though recognition of objects proceeds by the means of a schema, an abstract form shared by all members of a certain kind, each particular image also differs from others of its kind. That is, they differ in the additional features which are not determined (entailed) by the concept. For instance, I recognize the flower by the application of the flower rule (schema) to the sensible manifold. The flower rule is a basic figurative mental representation of an object with petals, leaves and stems in a specific relation. Yet, a particular image of a flower may have a distinct shape of petals in a particular combination of colours (that is, a distinctive combination of the general features). But these distinctive features of this particular flower are not entailed by the concept of a flower.[5] This is because, as Kant writes, a concept 'refers to the object indirectly, by means of a characteristic that may be common to several things' (A320/B377). That is to say, concepts can serve as rules only for the features of the object common to members of a certain kind, but cannot be rules for the individual features and their combinations which are distinct and unique to the particular object itself. As Sarah Gibbons in her analysis of Kant's imagination puts it: 'Concepts can only provide a discursive unity of diverse representations possessing some common feature; they do not represent those diverse representations as parts of a single encompassing whole' (1994: 44).[6] Thus, the presence of these additional features which are not entailed by the concept shows that the activity of imagination is not fully determined by the concept, and therefore can potentially be in free play.

My suggestion is that aesthetic attributes refer to those specific and distinctive features of an object that have been left undetermined by the concept of the object.[7] Such an interpretation appears to be suggested by Kant in his discussion of aesthetic ideas where he writes the following:

> In the use of the imagination for cognition, the imagination is under the constraint of the understanding and is subject to the limitation of being

adequate to its concept; in an aesthetic respect, however, the imagination is free to provide, beyond that concord with the concept, unsought extensive undeveloped material for the understanding, of which the latter took no regard in its concept, but which it applies, not so much objectively, for cognition, as subjectively, for the animation of the cognitive powers.

(KU 5:317; 194)

As I read this passage, the 'unsought extensive undeveloped material' that sets imagination and understanding into a free play refers to aesthetic attributes that furnish an aesthetic idea. Kant explains this undeveloped material as something that has been left out by the ordinary synthesis according to a determinate concept. That is, the object has been recognized as the particular kind of thing; it is, say, a flower; yet, the sensible manifold in this object contains an additional material that is not determined by the concept of the object. It is this additional material of the object that aesthetic attributes take into account. To illustrate my proposal, I will consider Picasso's *Weeping Woman* (1937).

The painting depicts a grief-stricken face of a woman, weeping into a handkerchief. One can immediately recognize that this is a painting of a human face. One perceives the head, eyes, nose and lips (that is, logical attributes), as presented by the schema of a human face. But one also perceives a specific and distinctive configuration of these features. The face is painted in different geometrical shapes, split into fragments; the shapes of the mouth, teeth, tears and the handkerchief used to dry the tears are almost fused into each other; the sides of the face are juxtaposed in such a way that they offer simultaneously a frontal and a profile perspective of the face. But these distinctive features are not specified by the schema of a human face. Hence, they are products of imagination in its free play.

Yet, one can also notice that it is precisely these distinctive features of a face, combined together in this specific way, that bring to mind the thoughts, feelings, moods, sensations and other mental representations associated with the idea we generally have of grief and loss. For instance, the feeling of disorientation and anxiety as occasioned by the simultaneous presence of the frontal and profile perspectives of the face, the idea of emotional brokenness conveyed by the fractured shapes of the face, brought together in an unnerving way or the idea of a paralysing emptiness as beautifully expressed through the image of woman's chattering teeth, convulsively nibbling the handkerchief. It is the imaginative synthesis of these distinctive features of a woman's face which bring to mind different associational thoughts (i.e. aesthetic attributes) connected with the idea we have of grief or loss. Even though these ideas do not have a determinate empirical counterpart, they can nevertheless be depicted through the synthesis of aesthetic attributes (a collection of associative images). In this way we are able to think about these ideas in terms of a concrete perceptual experience.

Literary art expresses aesthetic ideas similarly, even though it does not speak to us by means of direct sensible presentations (or images) as visual art, but rather by means of words and concepts. While literary art communicates aesthetic ideas through linguistic elements, what sparks the animation of aesthetic attributes is not the determinate content of words and concepts, but rather additional mental representations evoked by the imagination. As Kant writes: 'poetry and oratory also derive the spirit which animates their works solely from the aesthetic attributes of the objects, which go alongside the logical ones, and give the imagination an impetus to think more, although in an undeveloped way, than can be comprehended in a concept, and hence in a determinate linguistic expression' (KU 5:315; 193). Kant illustrates his idea with the following poem written by Frederick the Great:

> Laßt uns aus dem Leben ohne Murren weichen und ohne etwas zu bedauern, indem wir die Welt noch alsdann mit Wohltaten überhäuft zurücklassen. So verbreitet die Sonne, nachdem sie ihren Tageslauf vollendet hat, noch ein mildes Licht am Himmel; und die letzten Strahlen, die sie in die Lüfte schickt, sind ihre letzten Seufzer für das Wohl der Welt.
> (1790, cit. from Kant's gesammelte Schriften, vol. V, Berlin: Druck und Verlag von Georg Reimer, 1908, pp. 315–16)

> Let us depart from life without grumbling and without regret, leaving the world behind us lavished with good deeds. Thus does the sun, having completed its daily course, still spreads a mild light in the heaven; and the last rays it sends into the sky are its last sighs for the well-being of the world.

Presumably, the poem animates the rational idea of the cosmopolitan disposition towards death by means of 'an attribute that the imagination (in the recollection of everything agreeable in a beautiful summer day, drawn to a close, which a bright evening calls to mind) associates with that representation' (KU 5:316; 194). Aesthetic attributes are constituted not by the conceptual meaning of the summer day (as a long, warm and pleasant day in a summer), but rather by the particular mental representations or mental images of a summer day that one has previously experienced and associated with various feelings, thoughts and sensations. Just like the evening summer sun takes its final bow after nourishing the earth so too does a person with a cosmopolitan attitude depart from his own life with peaceful, remorseless thoughts and resolved feelings. It is important to point out here that the content of aesthetic attributes consists not only from mental associations that are culturally determined (such as the association of the summer day with ideas of hopefulness, freedom and warmth), but also by personally idiosyncratic associations that one brings to mind when recollecting their own experience of a beautiful summer day. The content of our experience of aesthetic ideas always involves, as Jenny McMahon nicely points out, 'the fragments, feelings, intimations and other associations that one has inadvertently accumulated over one's lifetime' (2010: 435) and which we relate to the particular abstract concept or rational idea instigated by the artistic expression of aesthetic idea. After all, Kant describes aesthetic ideas as indeterminate in nature and as 'opening up for it the prospect of an immeasurable field of related representations' (KU 5:315; 193). The meaning of an aesthetic idea is inexhaustible, generating multifarious interpretations as they are constituted by the associational thoughts (or aesthetic attributes) rich with personal semantic memories, images, feelings and emotions.

Kant writes that in contrast to an empirical intuition, which is an external representation of imagination, an aesthetic idea is an 'inner intuition' of imagination (KU 5:314; 192). Although he does not explain what he means by such inner intuitions, taking into account his remark that an aesthetic idea is a 'coherent whole of an unutterable fullness of thought' (KU 5:329; 206), it is reasonable to assume that an aesthetic idea is a kind of a holistic mental representation (or a pattern) of various semantic or intellectual elements combined and unified together. A similar suggestion is also given by Andrew Chignell who interprets an aesthetic idea as a 'plurality of representations or thoughts linked together' (2007: 424). Jennifer McMahon appears to agree as she describes an aesthetic idea as a 'loose collection or constellation of fragments, feelings, ideas, and facts brought together merely by being associated with a concept, though not unified or subsumed under it' (2017: 436). We can conclude that an aesthetic idea refers to a certain kind of inner picturing of thoughts and associations that occur in our mind as we reflect on a particular object or an artwork and which gives rise to ideas that go beyond sensory experience.[8]

We can see that the relationship between aesthetic ideas, aesthetic attributes and ideas that go beyond sensory experience (that is, rational ideas and abstract concepts) is analogous to the relationship that exists between empirical intuitions, logical attributes and determinate concepts. Just like an empirical intuition (say, an image of a flower)[9] is a concrete representation of logical attributes (set of marks thought in the concept of a flower) that constitute the determinate concept of a flower, so too an aesthetic idea appears to be a concrete representation of aesthetic attributes that constitute the thoughts and experiences we have of ideas that go beyond sensory experience (say, of the idea of grief). That is, both empirical intuition (say, flower-empirical intuition) and an aesthetic idea (say, grief-inner intuition) are particular representations of imagination; they are both product of the synthesis of imagination. But whereas an empirical intuition is a product of the imaginative synthesis of various sense impressions, an aesthetic idea, as an inner (mental) picture, is a product of the imaginative synthesis of various thoughts and associations (that is, aesthetic attributes).

Accordingly, the formation of an aesthetic idea appears to be similar to the formation of empirical intuition in that it also requires the harmonious activity between the faculties of imagination and understanding. In short, it is Kant's idea that we come to form an image by means of the faculty of imagination, which is responsible for combining together the manifold of intuition given to us by the receptive faculty of sensibility, and the faculty of understanding, which unifies this manifold under the concept of the object. As Kant claims, even though the sensibility provides us with the manifold of intuitions (understood in the narrow sense as un-synthesized sense impressions), this manifold is indiscriminate when received through the senses, because different representations within the manifold are 'dispersed and separate in the mind' (A120). The manifold of intuition becomes intelligible only through the imaginative synthesis: 'Although intuition offers a manifold, yet intuition can never bring this manifold about as a manifold, and as contained moreover in one presentation, unless a synthesis occurs in this process' (A99). This synthesis is performed by the faculty of imagination and it is called apprehension: 'For the imagination is to bring the manifold of intuition to an image; hence it must beforehand take the impressions up into its activity, i.e., apprehend them' (A120). To produce an image (say, an image of a dog), the imagination must gather together the manifold of intuitions (say, barking-intuition, head, four legs, tail, fur-intuitions), that is, it must synthesize them. We need to combine together all the partial representations present in the manifold in order to have a complete and unified representation of an object.

In the formation of an empirical intuition, the imaginative synthesis is dependent on determinate concepts and on the faculty of understanding as the 'faculty for thinking of objects of sensible intuition' (A51/B75). Without the faculty of understanding and its concepts we would not be able to think or understand intuitions given through sensibility. To bring intuitions to concept is to make them 'understandable' (A51/B75). Kant has a twofold definition of concepts. On the one hand, a concept is 'this one consciousness [that] unites in one presentation what is manifold, intuited little by little, and then also reproduced' (A103). We attain conceptual consciousness when we recognize the synthesized sense impressions as falling under a particular determinate concept. Only when we have applied the determinate concept to the various intuitive representations (for example, applying determinate concept of a dog to the various intuitive representations of a dog that we have apprehended and reproduced) can we attain the consciousness of what it is that we are perceiving (that is, the image of a dog). On the other hand, a concept is not merely that which provides consciousness of what it is that we are perceiving; rather, it also functions as a rule: 'A concept, in terms of its form, is always something that is universal and that serves as a rule' (A106). That is to say, a concept serves as a rule for the synthetic unity of a manifold of intuition: 'This unity is impossible, however, unless the intuition can be produced according to a rule' (A 105). In other words, a concept is responsible not only for distinguishing different sense impressions from each other, for example, that dog-intuitions such as perceiving tail, head, four legs all belong together, but also for organizing these dog-intuitions in a specific way. In order to have an image of a dog it is not sufficient only to have an awareness of the content (matter) of the concept (for example, that a concept of a dog is a set of marks such as animal, four legs, fur and barking). Rather, what is also required is to organize these sense impressions in a specific way as governed by the conceptual rule. The concept serves as a plan for the synthesis of sensible manifold, determining the way we come to construct the image.[10]

The formation of an aesthetic idea proceeds similarly in that it also involves the synthetic activity of the faculty of imagination. An aesthetic idea is comprised from the manifold of aesthetic attributes, that is, a plurality of associational thoughts. Each aesthetic attribute carries forward a particular associational thought; yet, it is their combination that can communicate a coherent and unified meaning. That is to say, if aesthetic attributes are not synthesized together by the faculty of imagination, then they are perceived as separate and dispersed thoughts, without any connection to each other and thus without any intelligible meaning. The imagination must gather and combine together the manifold of aesthetic attributes, while understanding must provide order and unity to it. However, while in empirical intuition-formation the imagination synthesizes the sensible manifold as specified by the determinate concept of the understanding (that is, in order to recognize a particular object, say as a flower, the imagination must follow the rule as specified by the concept of a flower), in the aesthetic idea-formation the synthesis of imagination is not governed by any determinate concept and thus it is in a free play: 'The aesthetic idea can be called an inexponible representation of the imagination (in its free play)' (KU 5:343; 218). In other words, there is no determinate concept guiding the imagination as to how it ought to combine the manifold of aesthetic attributes as it is the case in ordinary perception, where the sensible manifold is organized according to the conceptual rule.

It is important to point out that even though the imaginative synthesis of aesthetic attributes is not governed by any determinate concepts, it is not completely free of concepts. As mentioned previously, aesthetic ideas are indirect sensible representations of concepts that go beyond sensory experience, such as rational ideas and abstract concepts. It is these kinds of concepts that serve as rules guiding the synthesis of imagination in the formation of an aesthetic idea. Yet, it is distinctive for such concepts that no determinate rules can be given for them, moreover, that they have a significantly subjective nature in that they are dependent on our own subjective experiences. For example, an artist's intention may be to represent an idea of grief or idea of loneliness and complexity of human existence. Yet, one does not know what these ideas ought to look like, that is, one does not have an appropriate empirical intuition for such ideas (in comparison to the empirical intuition of, say, a dog). But if one does not have an empirical intuition for such ideas, then one does not have determinate rules in accordance with which to produce (or recognize) a manifold for such ideas. Yet, this is the meaning of Kant's concept of the free play between imagination and understanding: 'The powers of cognition that are set into play by this representation are hereby in a free play, since no determinate concept restricts them to a particular rule of cognition' (KU 5:217; 102). Accordingly, what is distinctive for artistic expression of an aesthetic idea is that in spite of its dependence on the concept (this concept being rational idea or abstract concept), the aesthetic attributes that constitute an aesthetic idea are combined together freely, without being governed by any determinate rule as to how the combination ought to proceed.

How exactly imagination produces aesthetic ideas is difficult to say since Kant does not provide any explanation of such an imaginative operation. Presumably, the imagination is able to generate aesthetic ideas without being governed by any determinate concepts, while, however, being in accordance with the general need of the understanding to bring order and unity to the sensible manifold. Kant describes the ability to produce aesthetic ideas as 'the inborn predisposition of the mind (ingenium)' (KU 5:307; 186) and as a 'particular spirit given to a person at birth, which protects and guides him, and from whose inspiration those original ideas stem' (KU 5:308; 187). The artist himself cannot explain how he comes to produce aesthetic ideas, nor can he describe these ideas to others through the use of a determinate language. Kant writes:

> The author of a product that he owes to his genius does not know himself how the ideas for it come to him, and also does not have it in his power to think up such things at will or according to plan, and to communicate to others precepts that would put them in a position to produce similar products.
>
> (KU 5:308; 187)

An aesthetic idea is conceptually indeterminate; it cannot be specified in the set of criteria that others could follow (for otherwise the product could not be an object of aesthetic appraisal). Kant claims: 'The rule must be abstracted from the deed, i.e. from the product' (KU 5:309; 188). That is to say, an aesthetic idea can be grasped and communicated to others only through direct observation and reflection on the particular work itself; merely hearing or reading someone else's description of the work cannot reveal the aesthetic idea.

This is compatible with my interpretation of an aesthetic idea as generated by the unique and distinct features of an object (aesthetic attributes), in contrast to those general features that the object shares with others of its kind and can be explicitly articulated (logical attributes). We can, for example, explicitly articulate criteria for why we would classify something as a face, or as a flower, without having to be directly acquainted with the object itself. Yet we cannot state such criteria that uniquely identify particular objects in all their detail. For instance, it is impossible to give a description that would apply completely accurately and uniquely to Picasso's *Weeping Woman*, and yet this particular work of art expresses an aesthetic idea. Now, if an aesthetic idea is produced by the synthesis of particular and distinct features of an object, but these particular features cannot be explicitly articulated, that is, one cannot completely describe all the features of the particular object (they can only be distinguished by observation), then it also follows that an aesthetic idea itself cannot be grasped in a determinate concept. As we have seen, concepts can only be based on commonalities between distinct particular objects but cannot represent the individual features of the object. The only way an aesthetic idea can be revealed and communicated to others is through a non-discursive, affective mode of expression. The communication of an aesthetic idea is subjective (as the term 'aesthetic' suggests) in the sense that it can be communicated only through a direct feeling of the mental state of free harmony between imagination and understanding. In other words, an aesthetic idea can be grasped only through an aesthetic experience of the work, that is, experiencing a sense of freedom and harmony in the playful interaction between different associational thoughts that the imagination conjures up without being governed by any determinate concepts, but which is nevertheless in agreement with the understanding – namely, the sensible manifold is organized in a way that exhibits indeterminate conceptual content of rational ideas and abstract concepts. Thus, even though Kant writes that an aesthetic idea is ineffable (cannot be expressed in a determinate language), it is nonetheless communicable, that is, it can be revealed and communicated to others through the aesthetic feeling of pleasure.[11]

Aesthetic ideas and cognitive processing of abstract concepts

I have so far argued that an aesthetic idea is similar to an empirical intuition in that they are both particular representations of imagination, that is, they are both product of the synthesis of imagination. They differ, however, with respect to the constituents of the syntheses. While in the case of an empirical intuition the synthesis is comprised of perceptual features (such as leaves, stem and petals forming the flower-intuition), in the case of an aesthetic idea the synthesis is comprised of aesthetic attributes, that is, thoughts, associations and other mental representations evoked by the distinctive features of an object or an artwork. If my account is correct, then it is fair to say that aesthetic ideas, as expressed in works of art, can capture and bring together various introspective, emotional and affective properties (that is, properties expressing our subjective experiences) that appear to be central to the content of abstract phenomena. Consequently, aesthetic ideas can help us to overcome cognitive limitations that we often experience in our attempts to articulate the meaning of abstract concepts. Before developing my argument, I will give a brief overview of the issues related to the apprehension of meaning and cognitive processing of abstract representations that have been of interest to cognitive science. This is a problem about the question of how abstract concepts are represented considering they do not have easily identifiable and clearly perceivable referents as concrete concepts have.

Most of us share the intuitive feeling that abstract concepts are harder to understand than concrete, empirical concepts.[12] This is evident, for example, from the feelings of insecurity and struggle that we experience each time we try to explain the meaning of concepts such as truth, love, hopelessness or vulnerability.[13] Ralph Ellis nicely describes this experience: 'When we begin to say what we mean by "in love", most of us find ourselves struggling, questioning and revising what we think we mean by it [...] There was an unsureness, a hesitance, a fear of saying what we did not mean, or not being able to say what we did mean' (1995: 73). Yet we do not experience any difficulty in grasping the meaning of concrete, empirical concepts. The meaning of a concept, say, the concept of a flower, is quickly available to us in terms of a specific set of physical properties that can be perceived by one or more of the senses. Ellis writes that our attempts in explaining the meaning of empirical concepts are accompanied by the feeling of confidence 'that we could call up certain images, but normally without actually calling them up in order to prove to ourselves that we can do it' (1995: 73). Our understanding of concepts is ultimately dependent on our ability to explain their meanings in imaginable terms. In comparison with empirical concepts, however, abstract concepts do not have a physical, perceptual and thus imaginable referent. That is to say, we cannot experience them directly through our senses. To the extent that such concepts lack a direct perceptual and imaginable counterpart, they are more difficult to comprehend and understand.

Such a view is supported by contemporary studies in cognitive science. Numerous research studies show that abstract concepts are much more difficult to understand than concrete or determinate concepts.[14] This difference is known as the concreteness effect and is commonly explained by two main theories – Dual Coding Theory (DCT) and Context Availability Theory (CAT). In short, DCT claims that comprehension of concepts depends on two interconnected systems: a verbal system, responsible for processing verbal information, and an imaginal system, responsible for processing non-verbal information and for generating mental images (Paivio 1990). According to this theory, abstract concepts are harder to understand because they lack an additional perceptual source of information that concrete concepts have. CAT, on the other hand, claims that conceptual processing depends on contextual and situational information (Scwanenflugel 1991). For example, understanding the meaning of the concept 'chair' depends not merely on knowing its physical properties, but also on relevant situations in which the object occurs or is used. Abstract concepts are more difficult to understand because they have a weaker connection to contextual information.

In sum, both theories show that perceptual information plays an important role in conceptual processing; in order for us to comprehend and fully understand abstract thoughts and ideas, they must in some sense be connected with concrete and imaginable representations. One way to evoke imagery for abstract concepts is by means of their associations with empirical concepts. For instance, the abstract concept 'religion' can evoke imagery indirectly by means of its association with the empirical concept 'church'. Or the concept 'justice' can evoke imagery through its association with a particular situation, such as a court trial. Availability of such referential imagery presumably eliminates the concreteness effect and produces better processing and understanding of abstract phenomena.

It is difficult, however, to see how such referential imagery could convey the meaning of abstract concepts as it is determined by our own subjective experiences. Numerous studies in contemporary cognitive science show that the content of abstract concepts involves not merely features that can be explicitly articulated in words and propositions, but also experience-related properties that are more difficult to express in ordinary language (Schwanenflugel 1991, Wiemer-Hastings and Xu 2005, Kousta et al. 2011, Barsalou and Wiemer-Hastings 2015, Kiefer and Harpaintner 2020). As Lawrence Barsalou and Katja Wiemer-Hastings point out, 'direct experience of abstract concepts appears central to their content' (2015: 133). Consider, for example, the concept of hopelessness. One can explicitly articulate the meaning of hopelessness as being an emotion with a negative view of the future. However, this conceptual meaning cannot give a full account of the idea of hopelessness as we experience it from the inside, as determined by our own beliefs, thoughts, memories, desires, feelings, interests, goals and so forth. That is to say, the conceptual meaning cannot fully capture detailed features of one's inner experience of hopelessness – for example, how it feels to have negative thoughts and beliefs about future, the interplay of feelings of powerlessness, worthlessness and incapability, and how these feelings can ultimately result in self-destructive thoughts and behaviours. When we try to explain the meaning of the idea of hopelessness, we feel that there is always much more that is implicitly present in our awareness of the meaning than we are able to explicitly articulate, and that this implicit meaning we internally experience is far more specific, precise and complex than we can capture in words. This is because, as Eugene Gendlin points out, '[n]one of our theoretical concepts are nearly specific and complex enough to come even close to the facets one feels. Logic and theory merely reconstruct some aspects of experience into a general pattern' (1968: 217). Our ordinary words and concepts refer to our experiences indirectly by means of general characteristics that are abstracted from our experiences, and as such they are logically incapable of grasping and communicating introspective, affective and emotional information associated with concepts such as hopelessness. Concepts are abstract constructs that select out of the experience merely some general patterns, thereby leaving behind a great deal of implicit feelings and thoughts associated with our concepts. Thus, what we ordinarily explicate by saying that we feel hopelessness is merely some rough aspect of our experience, while a great deal of meaning is left unelaborated and as such not fully comprehensible.

But how we experience hopelessness does appear to have an effect on our understanding of the concept of hopelessness itself. This is the idea emphasized by the contemporary theories of grounded cognition, which claim that abstract concept representations are essentially rooted in experiential information, such as introspection, emotional and affective states. This indicates that establishing and understanding the meaning of abstract concepts ultimately supervene on comprehending the internal relationship between various experiential information that constitute to a large extent the semantic content of abstract concepts. Mark Johnson, who argues for the qualitative dimension of meaning as part of our understanding of abstract phenomena, captures this idea accordingly: 'The meaning is in what you think and feel and do, and it lies in recurring qualities, patterns, and structures of experience that are, for the most part, unconsciously and automatically shaping how you understand, how you choose, and how you express yourself' (2007: 61). Something similar is pointed out by Eugene T. Gendlin, an advocate of the theory of experienced or felt meanings: 'Every individual lives in his subjective experiencing and looks out at the world from it and through it' (1997: 228). Both authors argue in favour of the idea that there is an additional meaning to our abstract concepts that goes beyond their logical meanings. While logical meanings of abstract concepts represent merely some general patterns of our experiences (say, that 'hopelessness' is an emotion with a negative view of the future), they cannot grasp all the details of our experience of hopelessness. That is to say, they cannot capture all the phenomenological and qualitative aspects associated with our abstract concepts. Accordingly, if it is true that the content of our abstract concepts also entails experience-related properties, then it would seem to follow that we cannot fully determine the meaning of abstract concepts until we include some of our subjective experiences as well. These experiences provide an additional source of information that is required for a more complete understanding of abstract phenomena.

Every subjective experience expresses one's relation to the world; it is shaped by one's conscious or unconscious beliefs, thoughts, feelings, motives and desires. Thus, what we directly experience is never just some internal affective state, such as anger or sadness, but rather something much more complex and distinct than can be fully carried forward by the word angry or sad. What we feel, as Eugene Gendlin points out, is the whole 'detailed person-situation complexities' and which incorporates our own interpretative aspects – how we come to construe or perceive the situation based on our implicit beliefs, motives, desires, goals, emotions and past experiences (1968: 209). Experiencing is always interactive: 'It is never just "in us." It is always at, or about, or for, or in a context of perceptions and events' (Gendlin 1965: 46). These implicit situational and interpretative aspects are intensely felt in our experience, even though they are not conceptually clear and distinctly delineated. They are part of our experiences and an experience has a meaning that it has in virtue of these aspects and connections that obtain between them. Thus, to understand the meaning of our experience requires us to comprehend how different introspective, affective and emotional information that constitutes the background of our experience is related to each other in terms of cause and effect, part and whole. Robert Solomon, an advocate of a cognitive theory of emotions, expresses this idea similarly: 'Emotional experience is unintelligible without understanding the background of an emotion' (2007: 241).

However, while we are mostly aware of the experience itself (say, that we feel angry, lonely or hopeless), we are not fully aware of the various introspective and emotional aspects that comprise the background of our experience. As Andrew Ortony nicely puts it, 'Experience does not arrive in little discrete packets, but flows, leading us imperceptibly from one state to another' (1975: 46). That is, the introspective and emotional aspects of an experience are often fleeting, evading and difficult to comprehend in all their details. Unlike ordinary perception of mind-independent objects, say seeing a table, whereby we can comprehend all the details of the object as long as we sufficiently long look at it, the state of our mental processes is continuously changing, which makes is difficult for us to comprehend all the passing mental processes and to grasp all the details of the experience. It is hard, if not impossible, to pay attention and to describe accurately all the movements and sequences of our thoughts and feelings, to identify, specify and differentiate various aspects of our experience. When we are in a state of an intense experience, we usually do not have the capacity to concentrate and to follow all the undergoing mental processes. Besides, it is argued that self-observation itself can often disrupt the genuine character of the experience (Marres 1989: 65–8). For example, if we concentrate on observing and analysing our experience while having it (say of anxiety, or anger), then our own introspection might change or diminish the intensity of the experience itself. This is because introspection requires focused attention and clarity of mind, but which is not something that is present in one's experience of anxiety or anger. As Eugene Gendlin verbalizes this point, 'One cannot expect to grasp clearly what the trouble is while it troubles' (1968: 222). Accordingly, we can see that it is difficult, if not impossible, to comprehend all the introspective and emotional aspects of our experiences. But not being able to comprehend all the details of our experiences makes the emotion and abstract concepts that refer to those experiences to certain extent cognitively unavailable.

My aim in the next section is to demonstrate that Kant's aesthetic ideas, as expressed in works of art, have a cognitive dimension in that they can reveal various relationships between different introspective, emotional and affective information that appear to be central to the content of abstract phenomena. Consequently, they can promote (objectual) understanding of rational ideas and abstract concepts, and thereby help us to overcome cognitive limitations that we often experience in our attempt to articulate them.

Aesthetic ideas and objectual understanding

As we have seen, our ordinary conceptual vocabulary is inadequate to fully communicate all the implicit meanings of our concepts such as love, hopelessness or freedom. The only way to express what is ineffable in our abstract concepts is by means of their associations with aesthetic attributes, that is, supplementary representations of imagination that express implications and kinships between different concepts and objects (say, kinship between the image of the fractured shape of a face and the idea of emotional brokenness in Picasso's painting) as only such imaginative representation are precise and concrete enough to capture the subtlety of thoughts and the nuances of felt meanings associated with our concepts. In other words, aesthetic ideas are concrete representations holding together various introspective, emotional and affective information involved in our experience of abstract concepts. The availability of such imaginary representations can profoundly expand the meaning of these concepts and further our understanding of them.

I believe this is the idea Kant has in mind when he writes that an aesthetic idea 'aesthetically enlarges the concept itself in an unbounded way' (KU 5:315; 193), thereby implying that expansion of a concept is not logical expansion that proceeds by adding actual properties or logical attributes to the concept.[15] Rather, the expansion of a concept proceeds aesthetically, that is, by means of aesthetic attributes that bring to mind a multitude of thoughts, feelings, moods and sensations connected with the given concept. One may, for example, notice that Picasso's *Weeping Woman* does not broaden our definition of grief as an emotional state experienced due to significant loss, but rather it stimulates our reflection on a whole range of reactions, emotions, thoughts, beliefs, other mental aspects and effects involved in our experience of grief and loss. Thus, the kind of understanding that aesthetic ideas add to our concepts refers to the comprehension of the various introspective, emotional and affective information that shape the meaning of our abstract concepts, but cannot be explicitly articulated in words and propositions.

This is at least how I interpret Kant's claims that an aesthetic idea 'occasions much thinking though without it being possible for any determinate thought, i.e., concept, to be adequate to it' (KU 5:314; 192), or 'gives more to think about than can be grasped and made distinct in it (although it does, to be sure, belong to the concept of the object)' (KU 5:315; 193); and 'no expression designating a determinate concept can be found for it, which therefore allows the addition to a concept of much that is unnameable' (KU 5:316; 194).

In these passages Kant uses words such as 'much thinking' or 'more to think' which seem to suggest that non-discursivity is caused by the great quantity of thoughts, which in principle could be articulated in a determinate language, but their excess makes it difficult to do so. This, I believe, is not what Kant has in mind. Thoughts provided by aesthetic ideas are not merely difficult; rather, they are impossible to conceptualize since they themselves are, as Kant puts it, 'unnameable'. The kind of thoughts that aesthetic ideas add to our concepts are such that they cannot be explicitly articulated in words and propositions, but, as Kant states, 'belong to the concept'. Yet the kind of thoughts that appear to meet these two characteristics – namely, non-discursivity and pertaining to the concept – are the kind of thoughts that refer to the introspective, affective and emotional aspects associated with our concepts.

Consider again the concept of grief. Even though we can explicitly articulate the meaning of grief as being an emotion, experienced due to the loss of someone or something important to us, this conceptual meaning cannot give a full account of the idea of grief as we experience it from the inside. As I have pointed out, our ordinary language is not rich enough or precise enough to grasp and explicate all the subtle aspects of our experience. What we ordinarily explicate by saying that we feel grief is merely some rough aspect of our experience, while a great deal of meaning is left unelaborated and as such not fully comprehended and understood. It is this unelaborated meaning of our abstract concepts that aesthetic ideas carry forward. Aesthetic ideas make experiential information involved in abstract concepts salient by connecting them with particular imaginable representations, thereby making them more cognitively accessible to us. They thus provide us with an additional source of information that is required for a more complete understanding of abstract phenomena.

Such a view is implied in Kant's claim that concepts without intuitions are empty (A51/B75). He refers to empirical concepts which need to be connected to empirical intuitions in order to make sense of experience. Empirical intuitions are required to establish objective reality of concepts and without them empirical concepts are mere words, lacking any substantive meaning. But the same can be said of abstract thoughts and ideas, such as the concepts of truth, love, grief, hopelessness and vulnerability which do not have a direct sensible counterpart and to this extent we cannot fully understand them. Only by connecting these kinds of concepts with sensible intuitions (by means of aesthetic attributes) can we truly say that we understand what they mean. Artistic expression of aesthetic ideas thus fills in the emptiness that abstract concepts on their own would leave without empirical intuitions.

The idea that aesthetic ideas can enrich and broaden the experiential dimension of our concepts has also been pointed out by Samantha Matherne in her claim that aesthetic ideas expand our subjective connections to concepts in that they reveal 'the richness of experience we too often overlook in the exigencies of everyday life' (2013: 30) and thereby provide us with ways of 'entertaining new possibilities or looking at the concept in different ways' (2013: 37).[16] She gives an example of Proust's description of a madeleine cake and writes: 'The aesthetic idea and aesthetic attributes involved in that passage augment my concept of a madeleine with subjective characteristics, like memory, childhood, and nostalgia' (2013). Although I am certainly in agreement with Matherne's view on the aesthetic idea as enlivening our abstract notions, I also find her account unsatisfying, for it does not explain how exactly aesthetic ideas come to expand our subjective connections to concepts.

My proposal is that artistic expression of aesthetic ideas expands our subjective connections to concepts and broadens our interpretation of experiences by expressing different meanings that rational ideas and abstract concepts can have for us. If aesthetic ideas are products of the imaginative synthesis of various thoughts and associations (that is, aesthetic attributes), this implies that they are able to capture and bring together various introspective, emotional and affective information involved in the content of abstract concepts and can thereby express the meaning of our concepts as it is determined by our own experiences. By bringing these aspects together into one unified whole, aesthetic ideas can help us grasp the connection between different aspects and thereby comprehend the meaning of an experience. As I pointed out previously, the meaning of our experience is always partially incomplete and becomes complete only after we grasp the interrelatedness of various mental aspects that comprise the background of our experiences. Aesthetic ideas furnish the meaning of an experience by selecting, specifying and bringing together different introspective, emotional and affective aspects. Since there can be many different ways of selecting and combining aspects, there can be many different meanings of an experience prompted by an aesthetic idea. Each expression of an aesthetic idea brings a different meaning of an experience and thereby a different perspective on abstract concepts. Herein lies the originality of aesthetic ideas. To see exactly how aesthetic ideas can express different meanings of our abstract concepts, let us consider Michael Haneke's film *The Seventh Continent* (1989).

The movie is an agonizing story of a well-situated Austrian family and their attempt to escape the feeling of emotional and social isolation in the modern world by choosing to commit a suicide. The mental state of emptiness and depersonalization that accompanies the everyday life of this family is represented through images of objects, rather than subjects. We do not see characters' faces, but merely fragmented and isolated shots of their hands turning off the alarm clock, opening curtains, putting toothpaste on a toothbrush, tying shoelaces, making coffee. Using a cinematic technique that emphasizes the state of imprisonment by our daily routines, Haneke manages here to give a perceptible form to the feeling of the emptiness of one's existence, and thereby provides us with a rare opportunity to recognize certain mental states, emotions and ideas that cannot be directly represented. Through the depiction of emotionless and depersonalized performances of our daily routines, the film represents the idea of emotional emptiness, that is, how these emotional states themselves look. We often experience such mental states, yet have difficulty in understanding it clearly. Through the objectification of the idea of emotional isolation, we have an extraordinary opportunity to perceive this emotion in a more formulated and comprehensive way. In particular, the film offers one of many possible ways to understand the experience of emotional emptiness and alienation. The meaning of an experience is brought out by carefully selecting and specifying certain experiential information. For example, the feeling of being trapped in a life of routine as expressed by the depiction of mechanically performed daily tasks, the idea of depersonalization and loss of communication as conveyed by the narration accentuating the monotony of the characters' day-to-day lives and their impersonal verbal exchanges, and how these feelings ultimately lead to the experience of despair and anger at the world, as vividly expressed by the images of the characters demolishing their house and all their possessions and finally their decision to escape the feeling of imprisonment by choosing to commit suicide. By bringing together different experiential aspects (introspective, emotional and affective) that shape the meaning of the concept of emotional isolation, the film allows us to perceive and grasp the internal relations between these aspects (how they fit together and interact with each other), thereby helping us to obtain an understanding of the concept of hopelessness itself.

Haneke's film offers one particular form that the idea of emotional emptiness and alienation can take, but there are many other possible ways of expressing this idea. To give an example, Edvard Munch's painting *Evening on Karl Johan Street* (1892) conveys the idea of emotional isolation and alienation by depicting a crowd of people, detached and isolated from one another, with indistinct faces. Thus, by emphasizing the experience of anonymity, isolation and loss of self-awareness, the painting adds to yet another meaning that the idea of alienation can have for us. Both Haneke and Munch have instantiated the same concept of alienation and emotional isolation (these concepts playing a role of a theme in an artwork); yet, they express a different meaning of an experience of alienation – a unique way as to how we come to think and experience these concepts.[17] That is to say, they express two different aesthetic ideas and communicate in two different ways the meaning of our abstract concepts as it is determined by our own subjective experiences. By bringing together various aesthetic attributes representing introspective, emotional and affective information that constitute the meaning of the idea of emotional isolation as we experience it from the inside, both artists provided us with a concrete imaginable representation that allows us to think about the idea of emotional isolation in a way linked to sensory experience.[18]

In sum, even though works of art as expressions of aesthetic ideas enlarge concepts merely aesthetically, they nevertheless have a cognitive function in that they contribute to 'the enlargement of the faculties' (KU 5:329) and serve as 'kinds of cognition' (KU 5:305; 184). Aesthetic ideas enlarge our cognitive faculties because they provide (indirectly, that is, by means of aesthetic attributes) a sensible counterpart to the ideas and concepts that cannot be directly exhibited in experience. In this respect they play a similar function as the image of a flower plays for the concept of a flower, namely they play the role of giving objects for cognition. Aesthetic ideas are valuable because they give us an insight into the world of ideas and state of affairs that lie beyond sensory experience. That is to say, they afford us with an opportunity to intuit and apprehend that which can never be fully presented by means of our sense. In this respect aesthetic ideas have a cognitive function in that they contribute to the improvement or deepening not of our ordinary cognition of the concept (that is, the identity conditions), but rather our understanding of it, and which involves, as Kirk Pillow states, 'the myriad ways we sort and compose, weigh and order the furniture of the worlds that we make by such means, all of which reflects our practical engagement with those worlds' (2001: 206). Aesthetic ideas, as expressed in works of art, are equipped with a rich and sophisticated vocabulary, a collection of associative images (aesthetic attributes) that can carry further the finely determined and more specific aspects of our experiences, which are more difficult to grasp and articulate by ordinary language. They can make us explicitly aware of what we formerly had only an implicit sense. Aesthetic ideas reveal the meaning of our concepts as it is determined by our own subjective experiences and thereby come to embody a form of understanding that has as its target not only the concept's identity conditions (say, what hopelessness or emotional isolation is), but rather its inner structure – recognizing all the experiential information (introspective, affective and emotional properties) and their relations that contribute to the meaning of abstract concepts and rational ideas. This is an important cognitive achievement that aesthetic ideas bring forth, even though it does not consist in factual truths and cannot be stated in a propositional form. Aesthetic ideas afford us with an opportunity to envision the various introspective, emotional and affective aspects connected with our concepts, thereby imbuing them with a more substantive meaning and understanding. Consider, for example, one's experience of hopelessness. While we may experience our own state of hopelessness, there are limits to the degree of understanding of the idea of hopelessness itself that is available only from our own states. Through expression of an aesthetic idea, however, we can gain a different perspective on this idea, for example, what the state of hopelessness might look like, which can consequently contribute to a richer understanding of this idea. Accordingly, aesthetic ideas, as expressed through works of art, make the experiential aspects involved in our concepts more cognitively accessible to us, by connecting them with concrete and imaginable representations and thereby providing an additional source of information that is required for a more complete understanding of abstract phenomena. Moreover, aesthetic ideas do not merely offer a better understanding of abstract phenomena, but by extension also understanding of our own experiences, thoughts, emotions, beliefs, desires and other mental aspects that constitute our own self-concepts. In other words, artworks as expressions of aesthetic ideas can also facilitate self-knowledge and self-development. To the explanation of this phenomenon, I turn in the next chapter. Before proceeding, however, I want to give a brief account of the much-discussed comparison between Kant's aesthetic ideas and metaphors.

Aesthetic ideas and metaphors

In contemporary discussions, Kant's aesthetic ideas are often identified with symbolic and metaphorical representations in terms of their creativity, inexhaustibility of meaning and cognitive status (Coleman 1974, Nuyen 1989, Pillow 2001, Forrester 2012). For example, A. T. Nuyen identifies an aesthetic idea with a symbolic presentation, which is 'the metaphorical process by which the imagination creates a metaphorical meaning' (1989: 95). On his account, aesthetic ideas are forms of symbolic presentations that function as metaphorical analogies between ideas and objects. He relies on Kant's explanation of symbolic presentations, which consist in comparing two apparently dissimilar, yet conceptually analogous things, such as hand mill and a despotic state. Even though there is no obvious similarity between the empirical object hand mill and the concept of despotic state, there is however a similarity 'between the rule for reflecting on both and their causality' (KU 5:352; 226). That is to say, the nature or definition of both concepts is similar in that they both display a similar structure in respect of being 'mechanical, insensible, and lacking in dynamism' (Nuyen 1989: 98). Given the similarity of the rules of reflection, the empirical object of hand mill can function as a symbol for the idea that cannot be empirically encountered, namely the despotic state. Nuyen argues that such symbolic presentation exhibits creativity in that 'similarity is not already existence or antecedently given' and thus the symbol maker 'creates a similarity between an idea and a sensible intuition' (Nuyen 1989: 99). Given the assumption of creativity that presumably underlies the symbolic presentation, Nuyen concludes that 'we can take Kant's account of the symbolic process as an account of the metaphorical process' (Nuyen 1989: 98).

Nuyen's interpretation of aesthetic ideas as forms of symbolic exhibitions has been heavily criticized by Kirk Pillow (2001) and Stephen Forrester (2012). They both rightly point out that aesthetic ideas significantly differ from symbolic exhibitions in that the latter depend on a pre-existent, yet hidden and non-obvious similarity between two different things and the symbol-maker must exercise his own productive imagination in discovering this similarity or bringing it into explicit awareness, whereas the former does not merely reveal, but rather create new similarities between two different phenomena. According to Kirk Pillow, Kant's text introduces two different conceptions of metaphors, namely, 'weak' conception of metaphor as analogies, which is represented in Kant's notion of symbolic exhibitions, and 'strong' conception of metaphor that he identifies with a more sophisticated interaction theory of metaphor and which takes place in artistic expression of aesthetic ideas. Symbolic exhibition is weakly creative because it consists merely in noticing or discovering already-existent commonalities between two different phenomena (such as hand mill and despotic state), whereas aesthetic ideas, as interaction metaphors, are highly creative as they 'establish original affinities between features of the world, rather than discovering pregiven ones' (Pillow 2001: 197). Interactionism describes metaphors as interactions between two unrelated phenomena, called primary and subsidiary subjects, where a set of associations or connotations is transferred from the subsidiary to primary subject and as a result our understanding of both subjects is transformed and improved (Black 1954–5). To borrow Max Black's example, metaphor 'man is a wolf' consists of the primary subject 'man' and subsidiary subject 'wolf'. When both subjects interact, a set of culturally and socially established beliefs we have about the subsidiary subject 'wolf' (beliefs such as that wolves are fierce, voracious, carnivorous, treacherous, etc.) is projected to the primary subject 'man'. As a result of such created similarities between the wolf and the man, the meaning of both subjects is extended providing thereby new perspectival insight on both subjects.

Given that both the metaphor-maker and the audience contribute to the ascription or creation of associational thoughts, metaphorical meaning necessarily exhibits semantic indeterminacy and inexhaustibility. The set of projected connotations includes not only culturally established meanings, but also finer and subtle differences called up by each hearer of the metaphor. Thus, metaphorical meaning cannot be specified by any determinate rule. As Kirk Pillow writes:

> A successful metaphor conveys a particular range of meaning in the array of relations it transfers, but it does not determine that meaning with finality, because the transfer varies across hearers and the set of relevant comparisons is in principle open to addition. Hence the creation of metaphorical meaning is interactive in the added sense that both speaker and recipient of the metaphor contribute to its meaning.
>
> (2001: 198)

According to Pillow, metaphor on interaction view bears a striking similarity to Kant's notion of aesthetic ideas in that they are both result of a highly creative process of inventing new set of inexhaustible associational thoughts between two different phenomena that cannot be grasped in a determinate concept or literal expression. For instance, Kant's typical example of the image of Jupiter's eagle with a lightning bolt in its talons, which expresses the aesthetic idea of the king of heaven, can be seen as a subsidiary subject that brings to mind a variety of shared connotations transferred to the primary subject, namely, the king of heaven. The interaction between Jupiter's eagle and the king of heaven, and the resulting associational thoughts generate a new understanding and a new way of interpreting the meaning of both subject. As Pillow concludes, artistic expression of aesthetic ideas 'brings together disjoin materials to express a new perspective that transforms our understanding of ourselves and the world' and it has this power because 'they express through metaphor in the interactionist sense' (2001: 202).

In my view, however, Kirk Pillow is too quick in dismissing the unique nature of aesthetic ideas in comparison to metaphors as he does not fully consider the role of free harmony and taste in the process of aesthetic idea formation. Aesthetic ideas bear similarity to metaphors in that they are constituted by the plurality of shared and personal connotations or associational thoughts, which evoke the inexhaustibility of meanings; yet, these created associations or similarities are not a sufficient condition to experience the expression of an aesthetic idea. While associational thoughts constitute the semantic content of aesthetic ideas, it is nevertheless their formal structure that brings faculties of imagination and understanding into a free harmonious play that generates the expression of aesthetic ideas, and which we experience aesthetically by means of the feeling of pleasure. It is not the content of aesthetic ideas, namely, associational thoughts and evoked similarities, but rather their coherent combination that necessitates the experience of an aesthetic idea and this is a matter of taste or aesthetic reflective judgement. Taste, Kant writes, is a 'regulative faculty of judging form in the combination of the manifold in the power of imagination' (Anthro 7:247; 143). In other words, taste is a critical capacity to experience a sense of freedom in the playful interaction between different associational thoughts that is not governed by any determinate rule, but which is nevertheless in harmony with the understanding in the sense that the formal structure of associational thoughts is appropriate for the expression of an aesthetic idea. We can see accordingly that Kant's notion of aesthetic ideas is essentially connected with his theory of beauty as a form of purposiveness without purpose (or free harmony). As I will argue later, in contrast to discursive recognition of metaphorical meanings, the communication of the semantic meaning conveyed by aesthetic ideas is grasped or recognized aesthetically, by means of the feeling of pleasure. An aesthetic idea is after all an intuition, an image of some sort (indirectly) representing concepts that go beyond sensory experience. The function of aesthetic ideas is not exhausted merely by seeking and creating similarities between different phenomena, which can in principle be discursively expressed; rather, their ultimate aim is to find intuitions for concepts that go beyond sensory experience and thereby provide 'kinds of cognition' (KU 5:305; 184) for them.[19]

3

Artistic expression of aesthetic ideas and therapeutic self-knowledge

I argued in the previous chapter that works of art, as expressions of aesthetic ideas, have a cognitive value in that they reveal the various relationships between the introspective, emotional and affective information that appear to be central to the semantic content of our rational ideas and abstract concepts, thereby promoting our understanding of them. These are the kind of concepts that refer to our mental states, emotions and personality traits and are to a large extent determined by our own beliefs, thoughts, feelings, motives, values, desires and other mental aspects that constitute our own self-concept. This suggests that works of art can also, by means of revealing the experiential dimension of our abstract concepts, give us the opportunity to recognize various mental and emotional aspects that constitute the background of our own experiences, thereby contributing to promote our own self-knowledge.

The idea of works of art serving a role of cultivating self-knowledge has been seldom discussed in contemporary aesthetics and philosophy of art. Philosophical and psychological debates are primarily concerned with understanding the socio-epistemic and moral effects of art, and with the importance of art engagement in cultivating empathy and our ability to understand others (Nussbaum 1990, Kieran 1996, Carroll 2002). Works of art, particularly of narrative kind, depict complex situations, motives and actions that give us the opportunity to imagine what it is like to be in another person's situation and thus can serve to promote our moral and social sensibilities with respect to the lives of others. Although these discussions may help explain how art can lead to reflection on our own norms, values and moral principles,[1] they do not say much about the effects that works of art have on promoting self-knowledge that is not socially and morally oriented, namely so called 'factual self-knowledge', which refers to knowledge of our own experiences, thoughts, emotions, beliefs, desires and other mental states that constitute the self.² While, for most of us, it seems obvious that art has these effects, little is known about how and why they occur. Addressing this issue is the main aim of this chapter. I intend to show that our engagement with works of art as expressions of aesthetic ideas and thus as (indirect) sensible representations of various introspective, affective and emotional aspects involved in our abstract concepts gives us a unique opportunity to adopt a dual (first- and third-person) perspective on the self. As it has been argued recently by psychologists and philosophers of mind, such a dual perspective is necessary for obtaining the kind of self-knowledge that leads to self-change, that is, therapeutic self-knowledge (McGeer 2008, Bell and Leite 2016, Strijbos and Jongepier 2018). It is distinctive for such self-knowledge that it does not provide us with completely new information about our own self; rather, it allows us fully to recognize various existent psychological aspects of our inner life of which we are often not completely aware and which guide our emotional experiences and behaviour (such as implicit beliefs, desires, feelings, motives and such).³ Self-knowledge in this respect functions as an act of recognition or knowing again, as Rita Felski aptly describes it: 'Something that may have been sensed in a vague, diffuse, or semi-conscious way now takes on a distinct shape, is amplified, heightened, or made newly visible. In a mobile interplay of exteriority and interiority, something that exists outside of me inspires a revised or altered sense of who I am' (2008: 25).

In short, to obtain therapeutic self-knowledge, we must relate to our own mental states from a first-person perspective, since this perspective generates the feeling of ownership and authority over our own mental life and can thus facilitate self-change. Yet we must also regulate first-person perspective by adopting a more distant, third-person perspective on our self, since the third-person view allows us to acknowledge the meaning of our experiences in the larger context of our life. While it is difficult to attain such a dual perspective on the self without psychotherapeutic interventions, I argue that works of art allow us to experience both perspectives, thereby helping us make sense of our own internal experiences and to recognize the meaning of these experiences in the larger context. To develop my argument, I begin with addressing the sources, values and limits of both the first- and third-personal conception of self-knowledge. I outline the principle of the first-person authority as being inherently connected with the first-personal conception of self-knowledge, one claimed to be a necessary condition for obtaining therapeutic self-knowledge. Next, I turn to the notion of art understood as a narrative simulation and describe two different types of emotional engagement with fictional characters that such simulation affords. Finally, I propose an account of the role of narrative simulation in promoting therapeutic self-knowledge.

The value and limits of self-knowledge

Self-knowledge refers to knowledge of our own mental states, such as our experiences, thoughts, emotions, beliefs, desires and other mental states that constitute the self (Gertler 2011). In general, to have self-knowledge is to have a clear sense of who we truly are, which is necessary for leading a meaningful and authentic life.

Literature usually points out two main sources of self-knowledge: introspection as the source of a first-personal self-knowledge and self-perception as the source of a third-personal self-knowledge (Wilson and Dunn 2004). Introspection refers to a direct awareness of our mental states. It relies on the inner information that we alone have regarding our mental life. For example, we know introspectively that we are angry by directly noticing the feeling of anger in us.[4] We have a first-personal relation to the content of self-information. This is different in the case of self-perception where we become aware of our own mental states indirectly, by means of inferring them from our behaviour (for instance, yelling and kicking being some of the behavioural signs indicating that we are angry). We have, in other words, a third-personal relationship towards our own mental states.

In contemporary philosophy of mind, introspection is no longer described in a Cartesian way as an inner perception of mental states that is similar to our perception of external objects (Taylor 1985, Moran 2001, Finkelstein 2003, Bilgrami 2006). The argument is that in contrast to external objects, mental objects are not independently existing facts waiting for our observation. Mental phenomena, as John McDowell points out, have no existence independently of our awareness of them (1996: 21). That is, without some awareness of, say, our own feeling of anger, we cannot say that we are angry. How we will come to conceive our own mental states will to a certain degree determine what these states will be for us. As David Finkelstein articulates this idea: 'Mental state self-ascriptions are unlike observation reports in that they constitute, to some extent, the facts to which they refer' (2003: 81). This implies that introspective awareness of our own mental states will have certain consequences on the nature of those mental states. In other words, introspection (partly) determines the nature of mental phenomena. Introspective awareness shapes the meaning of our mental states, because the process of bringing mental states into explicit awareness necessarily involves the activity of specifying, articulating and making sense of our mental states (Gendlin 1964, Varga 2015: 85–8). To know what we are thinking, feeling or experiencing essentially involves the activity of self-interpretation. Charles Taylor frames this idea well in the following passage:

> Built into the notion of representation in this view is the idea that representations are of independent objects. I frame a representation of something which is there independently of my depicting it, and which stands as a standard for this depiction. But when we look at a certain range of formulations which are crucial to human consciousness, the articulation of our human feelings, we can see that this does not hold. Formulating how we feel, or coming to adopt a new formulation, can frequently change how we feel. When I come to see that my feeling of guilt was false, or my feeling of love self-deluded, the emotions themselves are different [...] We could say that for these emotions, our understanding of them or the interpretations we accept are constitutive of the emotion. The understanding helps shape the emotion. And that is why the latter cannot be considered a fully independent object, and the traditional theory of consciousness as representation does not apply here.
>
> (1985: 100–1)

We have a special self-interpreting relationship towards our own mental states that does not apply to our observation of external objects. Our self-interpretations play a role in constituting the identity of our mental states. For this reason, introspection or relating to our own mental states from a first-personal perspective lies at the heart of psychotherapy (Bell and Leite 2016, Strijbos and Jongepier 2018). Psychotherapy depends on the assumption that we are in a privileged position of knowing our own mental states (what we are feeling, thinking, desiring) by simply experiencing these mental states. This privileged position towards knowing our own mind is reflected in the notion, known in contemporary philosophy of mind, as the principle of the first-person authority (Moran 2001, Finkelstein 2003, Bilgrami 2006). The principle is expressed in the idea that we are an authority on what we say (or think) that we feel, want, believe or desire – not merely in the sense of an epistemic authority, that is, having a privileged epistemic access to our own mental states (for example, our first-personal self-ascriptions of beliefs, thoughts and feelings are something that we take to be true, without taking into account any evidential considerations), but rather in the sense of an agential authority (Moran 2001, Parrott 2015). In short, we have an agential authority over our own mental states when we relate to our own state in the first-personal way through introspection. Richard Moran captures this idea:

> What is left out of the Spectator's view is the fact that I not only have a special access to someone's mental life, but that it is mine, expressive of my relation to the world, subject to my evaluation, correction, doubts, and tensions. This will mean that it is to be expected that a person's own awareness of his mental life will make for differences in the constitution of that mental life, differences that do not obtain with respect to one's awareness of other things or other people. For this reason, introspection is not to be thought of as a kind of light cast on a realm of inner objects, leaving them unaltered.
>
> (2001: 37)

The agential aspect of the first-person authority lies in our acknowledgement that psychological states are our own, and thus, that we are in a privileged position to do something with respect to those psychological states, that is, to shape them in a way that we find appropriate.

There are two ways of understanding the sense of appropriateness related to our own mental states. According to Moran's rational account, the sense of appropriateness lies in our ability to produce reasons as to what to feel, think or believe. As he writes:

> part of what it is to be a rational agent is to be able to subject one's attitudes to review in a way that makes a difference to what one's attitude is. One is an agent with respect to one's attitudes insofar as one orients oneself toward the question of one's beliefs by reflecting on what's true, or orients oneself toward the question of one's desires by reflecting on what's worthwhile or diverting or satisfying.
>
> (2001: 64)

For example, to answer the question whether it is appropriate to feel fear we must consider the reasons we have for holding this belief, that is, whether our assessment of the situation as threatening is really justified. Thus, we act as agents when we make rational deliberations as to what is appropriate, right, just or worthwhile to feel, think or believe. We have an agential authority insofar as we have rational authority. Our rational reflections determine the appropriateness of our mental states: 'When the articulation or interpretation of one's emotional state plays a role in the actual formation of that state, this will be because the interpretation is part of a deliberative inquiry about how to feel, how to respond' (2001: 58–9).

Moran develops his rational account of first-person authority primarily in the context of our moral deliberations; as such, it cannot be properly applied to our feelings and emotions that often remain unresponsive to rational deliberations (Carman 2003, McGeer 2008, Kloosterboer 2015, Strijbos and Jongepier 2018). For example, it is often the case that we feel happy without having any conscious reasons to believe we should feel happy. The opposite is also the case. Sometimes we have reasons to believe we should feel happy; yet, we cannot feel so. A nice illustration of the latter is given by Leo Tolstoy in his Confession, where he writes the following:

> I grew sick of life; some irresistible force was leading me to somehow get rid of it (…) And this was happening to me at a time when, from all indications, I should have been considered a completely happy man; this was when I was not yet fifty years old. I had a good, loving, and beloved wife, fine children, and a large estate that was growing and expanding without any effort on my part. More than ever before I was respected by friends and acquaintances, praised by strangers, and I could claim a certain renown without really deluding myself. Moreover, I was not physically and mentally unhealthy; on the contrary, I enjoyed a physical and mental vigor such as I had rarely encountered among others my age (…) And in such a state of affairs I came to a point where I could not live; and even though I feared death, I had to employ ruses against myself to keep from committing suicide.
>
> (1983: 28–9)

In spite of the fulfilment of his life goals and needs, Tolstoy cannot help but feel unhappiness and despair. Even though his self-ascription of unhappiness appears to conflict with his rationally endorsed reasons, it is nonetheless coming from a first-personal perspective. His unhappiness is a current perspective of the world that shapes his motivations, desires, thoughts and actions.[5]

Our emotions and feelings often remain unresponsive to rationally endorsed reasons because, as Naomi Kloosterboer argues, in contrast to reasons relevant in deliberating about belief-like states, which have to do with the truth-value of the content of the belief, reasons relevant in deliberating about emotions and feelings have to deal with our subjective relation to the world, namely, what matters to us, what we care about given our own conscious or unconscious beliefs, thoughts, feelings, motives, desires and values. As she points out: 'Our emotions are conceptually related to our concerns in the sense that they are responses to things that are of our concern' (2015: 252). Thus, Kloosterboer argues that, to answer the question whether it is appropriate to feel unhappy or angry, we must reflect on reasons as to why we come to evaluate the situation as dissatisfying or offensive, and this means that we must primarily answer the question as to who we are, what we value, desire and expect from others: 'Telling whether something is hurtful, offensive, or joyful for a specific person is grounded in considerations that depend upon who that person is, with certain character traits, concerns, plans, ambitions, fears, vulnerabilities, relations to other persons and so on' (2015: 253).

Our feelings of happiness, sadness or anger are appropriate insofar as they resonate with other aspects of our personality – our own beliefs, desires, goals, values and idea of the person we want to be. This is the sense of appropriateness towards which psychotherapy aims and which is reflected in the so-called affective account of the first-person authority, as recently proposed by Derek Strijbos and Fleur Jongepier (2018).[6] It relies on the basic rule of psychotherapy, namely, that we ought to give voice to our emotions and thoughts, that is, to articulate them and to attend closely what such self-articulations or self-interpretations 'stir up' in us, in the sense whether they resonate with other aspects of our personality. As Taylor Carman articulates this idea: 'The point is not to decide how I ought to feel, but to get clear about how I do feel by letting my emotions take shape and find a voice in what I say and do' (2003: 404). In other words, the acquisition of self-information must be 'emotionally convincing' rather than merely rationally apprehended (Abend 2007: 1437). The acquisition of self-information must be affectively apprehended in order to have a therapeutic effect. As Strijbos and Jongepier write, it is not the truth of the content of our feelings that is relevant here (what should I feel?) but rather the 'quality of one's emotion': 'if we find that a new description resonates with our emotions and makes them appear more clearly circumscribed and determinate, this will give us reason to believe that this new description is more accurate than the old one' (2018: 50). We come to feel, rather than rationally believe, the appropriateness of a specific self-articulation: 'It is enough that we come to experience the accuracy of our new descriptions on the basis of what these words stir up in us' (2018: 51). A given interpretation is constitutive of our mental state insofar as we find it emotionally convincing in the sense that it harmonizes with our own patterns of feelings, thoughts and actions, and makes these thoughts and feelings appear more focused, clear and coherent. We come to recognize emotionally convincing (appropriate) interpretation by the sense of relief, comfort and excitement it produces in us. We feel that all aspects of our experience are finally brought together and make sense. The result is an affective avowal, a full acknowledgement of our mental state. On the other hand, emotionally unconvincing (inappropriate) interpretation of our subjective experience provokes the feeling of discomfort, anxiety and lack of resolution. It is an interpretation that the individual experiences as inadequate or insufficient description of his own subjective experience. We experience this interpretation as less sense-making and disorganizing (that is, an affectively disavowed interpretation).

However, relating to our mental state from a first-personal point of view is not a sufficient condition for obtaining therapeutic self-knowledge because, as Strijbos and Jongepier write, 'when trying to find out what we feel, want, or believe, we cannot always trust our transparent outlook on the world' (2018: 51). Knowing our mind from a first-personal perspective is greatly vulnerable to epistemic fallibility and self-deception. As pointed out by numerous studies in cognitive science, we often form erroneous beliefs regarding our personality traits; misidentify motives and causes for our emotions, attitudes, choices and actions; and make wrong assessments regarding our dispositional mental states, such as moods, desires and beliefs.[7] Introspection appears to be epistemically unreliable source of self-knowledge due to the existence of various aspects of our inner life that remain outside of our conscious mind and have a significant effect on our cognitive and emotional experiences. Contemporary literature usually points out two main reasons as to why many of our mental states are inaccessible to direct consciousness (Wilson and Dunn 2004). First, because of motivational reasons to keep unpleasant thoughts, memories and feelings outside of awareness, either by means of the unconscious process of repression or conscious process of suppression (that is, intentional forgetting). In both cases the material continues to exist in our memory, influences our mental states and is in principle recoverable. Second, because most of our perceptual, semantic and affective processes are themselves unconscious, independently of any motivational reasons to keep unwanted mental states outside direct awareness. This idea is reflected in a current theory in cognitive science of the two distinct systems of information processing, that is, the unconscious system (implicit, impulsive or experiential system) and conscious system (explicit, rational or reflective system).[8] The conscious system, being evolutionary recent, is deliberate, slow, controlled and rule-based, associated primarily with language and reflective consciousness, while the unconscious system, on the other hand, is evolutionary older, non-verbal, associative, automatic, rapid and requires little cognitive effort and attention. The unconscious system plays an essential role in our mental life as it efficiently and quickly selects, interprets and organizes information. Seymour Epstein, one of the main proponents of the dual systems theory, emphasizes the importance of the unconscious system for our survival as it is responsible for performing daily activities that require quick actions. As he writes: 'without a system guided by affect, you might not even be able to decide whether you should cross the street' (2003: 161). Both systems are interactive and operate in parallel, which explains dissociations we often experience in our beliefs, attitudes, personality traits and feelings, as these mental states consist in dual information processing (Wilson, Lindsey and Schooler 2000).[9]

Since introspection refers merely to perceiving that what is immediately conscious to us and does not have a direct access to unconscious processes, it can often result in an incomplete and erroneous self-information.[10] Furthermore, introspection has been criticized on the ground that it leads to, what Jonathan Lear (2004) calls, the pathology of avowal, namely, to strengthening, rather than preventing, maladaptive mental states. In short, first-person perspective can often be governed by maladaptive implicit beliefs that are introspectively invisible to us. These implicit beliefs determine what features of ourselves and the world we will select and organize together, thereby preventing us from taking into account other features of ourselves and the world that might contribute to a more accurate self-understanding. For example, if a person holds the implicit belief that the world is full of betrayal (Lear's example), then this belief will guide his or her attention to those features of the world that are consistent with this belief while overlooking those that are not, thereby consequently reinforcing his or her implicit belief itself.

Thus, first-person perspective must also be regulated by adopting a third-person perspective on our own mental states. We must, as Victoria McGeer puts it, 'see ourselves as we see others – namely, as empirical subjects whose psychological states are responding to a variety of influences that are largely invisible from a naively egocentric first-person point of view' (2008: 101). The goal is to step back from our first-personal perspective, restrain ourselves from avowing, and re-access our inclinations. Third-person perspective refers to taking a spectator's stance towards our own mental states. According to a well-known version of this approach, the self-perception theory, individuals can acquire knowledge of their mental states by means of inferring them from observations of their own behaviour. Thus, as Daryl Bem writes, 'to the extent that internal cues are weak, ambiguous, or uninterpretable, the individual is functionally in the same position as an outside observer, an observer who must necessarily rely upon those same external cues to infer the individual's inner states' (1972: 2). Given that some of our behaviour is driven by implicit attitudes, motives and other psychological traits often unknown by us, observation of our own behavior can help us obtain a better insight into our internal states.

Third-person perspective, however, must be carefully regulated in order not to lead to self-alienation: namely, observing our own feelings, beliefs and desires as if they were someone else's involves experiencing our own mental states as something that happen to us, rather than being up to us.[11] If not regulated, third-person perspective can fail to take account of our awareness that the self under observation is ours and the importance this mode of self-awareness has for our self-development As Morris Eagle makes this point: 'Such knowledge may continue to have an impersonal "it" status, not necessarily in the sense of being ego-alien, but in the sense of not being fully integrated into one's first-person sense of who one is. One remains essentially the same person, having acquired additional information about oneself the way one would acquire information about a third person' (2011: 54).

To conclude this section, to acquire therapeutic self-knowledge, we must employ first-person perspective on the self since this perspective generates the feeling of ownership and authority over our own mental life. Yet, since first-person perspective can often be hijacked by psychological forces that are not directly conscious to us, it must also be regulated by adopting a third-person perspective on the self. That is, we must observe the meaning of our feelings and experiences from a broader perspective as they feature in the larger context of our life. Third-personally acquired insight into the psychological forces and impulses that lie beneath our immediate awareness is necessary for an accurate exercise of our agential authority.

My aim in what follows is to show that art as an expression of aesthetic ideas can allow us to experience both perspectives, thereby giving us a unique opportunity to acquire therapeutic self-knowledge. According to my account, works of art, particularly of narrative kind, offer different ways of interpreting our own subjective experiences from a third-person point of view, helping us obtain new information about our self. Furthermore, the process of acquiring self-information is not arbitrary; rather, it is bound up with the principle of the first-person authority. Considering that art is often involved in evoking emotions and feelings in us, the account of first-person authority applicable to art is the affective account; that is, the appropriateness of self-information is validated through the act of an affective avowal. I argue that we come to feel the appropriateness or inappropriateness of the self-information through the process of mental simulation, that is, imagining the depicted events and experiences from the point of view of characters. As pointed out previously, art as an expression of aesthetic ideas is a concrete sensible representation of the introspective, affective and emotional aspects that comprise our experiences. Works of art represent these mental aspects with more details, vividness and explicitness than we can ever encounter in our daily life. As John Gibson nicely points out, art is a dramatic achievement that presents our experience of the world 'not from an abstracted, external perspective but from the "inside" of life, in its full dramatic form' (2009: 482). Works of art can produce an immersive and absorbing experience that affects all our senses; they give us the opportunity to momentarily inhabit the world of characters and affectively engage with them. This allows us to feel more intensely the appropriateness (or inappropriateness) of acquired self-information for our own felt sense. Before proceeding to explicate my account in detail, I want to turn to the notion of narrative art functioning as a narrative simulation.

Narrative works of art as mental simulations

According to Raymond Mar and Keith Oatley, narrative works of art function primarily as 'simulations of selves in the social world' (2008: 173). There are two distinctive features of such narrative simulations. First, they give us an opportunity to put ourselves in the shoes of fictional characters, momentarily inhabit their thoughts, beliefs, desires and feelings. Narrative simulations encourage us to mentally simulate, that is, imagine mental states of characters from a first-personal perspective that we would have if we were actually experiencing portrayed events. Second, they offer an abstracted and simplified model of real-life experiences. Narrative simulation is a 'presentation of human relations and their outcomes in a compressed format' (2008: 183). That is, narrative simulation does not directly imitate our ordinary experiences, as they are too complex and detailed to be represented directly in an artform that is spatially and temporally limited. Rather, narrative simulation incorporates only those facts that are relevant for understanding the meaning of the story and the psychological situation of characters. For this reason, narrative art employs different strategies and techniques to convey information that cannot be explicitly given and thereby offers 'explanations of what goes on beneath the surface to generate observable behavior' (2008: 176). Each work of art, being a distinctive and particular expression of an aesthetic idea, carries a different explanation or interpretation of the portrayed issues and themes depending on the different ways of selecting, specifying and arranging elements of the story together. To see exactly how narrative art provides the interpretation of the depicted theme, let us consider again the Michael Haneke film *The Seventh Continent* (1989).

As mentioned in the previous chapter, the film offers one of many possible ways to understand the experience of emotional emptiness and alienation, namely, as one of the feeling of being imprisoned by the life of daily routines. For example, the film identifies and combines together specific introspective and emotional aspects connected with the idea we have of emotional isolation, such as the feeling of being trapped in mechanically performing daily routines, the feeling of depersonalization that characters exhibit in their daily communication with other people, their detached exchange of words, and the relationship between these feelings and their experience of disappointment, despair and anger at the world, as portrayed by the images of characters demolishing all their material possessions and ultimately their belief that the only way to escape the feeling of imprisonment is by choosing to commit suicide.

The film illustrates well in what way narrative art affords a dual (first- and third-personal) perspective on fictional characters.[12] On the one hand, the film offers a concrete and vivid vision of characters' psychological situation, which allows us to imagine the depicted event from the first-person perspective. In the film, we see characters' world through their eyes as they become emotionally desensitized, apathetic and despaired to the point that they decide to take their own lives. This creates a deeply felt simulating experience of their emotional states that gives us first-personal information about their inner experiences. On the other hand, the film also gives us the opportunity to obtain a more distant, third-person perspective of the portrayed events and characters, an effect accomplished by the formal and stylistic properties of the work. For example, close-up shots of characters mechanically performing their daily routines that exclude their faces or long takes depicting systematic destruction of all their possessions give us information about the psychological aspects that comprise the background of the family's emotional experience. For instance, that their feeling of emotional isolation is located in the particular way they come to interpret their daily routines and material possessions, namely, as something they are imprisoned by or confined in. The film thus provides additional information about the causal relationship between the characters' emotions and their psychological situation, which helps us to understand the destructive actions they take at the end. By giving us an explanation of characters' psychological and emotional aspects, the film invites us to shape a different relationship with these people, in which we are not merely participators in their story – experiencing their mental life through their eyes (first-person perspective) – but also external observers of their patterns of thoughts, feelings and actions (third-person perspective). Experiencing both first- and third-person perspective, interchanging during narrative simulation, gives us the opportunity to obtain more comprehensive understanding of characters and their psychological situation.[13]

Narrative art, however, does not merely give us an opportunity to adopt a dual perspective on the events and characters portrayed in the story, but, by extension, on our own psychological situation as well. That is to say, art simultaneously triggers reflection on our own experiences and personal characteristics. The explanation for this phenomenon, as I will show in more detail, lies in the nature of our own emotional responses to narrative works of art that reflect the workings of our emotions in general.

Therapeutic self-knowledge through art

As pointed out in the first section, our emotions reflect the sense of importance and worth we ascribe to something or someone and which is rooted in the particular way we are as a person. Each emotion we experience incorporates a sense of our own personal life concerns: what matters to us and what we care about, given our own personal characteristics, that is, our desires, goals, needs and beliefs about ourselves and the world we live in (Nussbaum 2004, Robinson 2005, Helm 2009). Personal life concerns (the sense of importance we ascribe to something or someone) differ from person to person given different personal characteristics; thus, the same act, situation or an event will not affect all of us in the same way. For instance, some feel pride upon professional success because they strive for achievements. Yet, those who do not share similar aspirations will not be affected by accomplishment in the same way. Similarly, some feel humiliated by insults because they have the need to be seen favourably by others. But others who do not care about opinions of others will not be bothered by such insults. We can come to know and understand each of our emotions by recognizing our own personal life concerns, and this means recognizing and identifying our personal characteristics.

Our emotional responses to fictional narratives are subjected to the same laws as emotions we experience in real life. This is the idea that follows directly from endorsing emotional realism, the view that we experience real, genuine emotions when engaging with fiction.[14] This means that how we come to engage emotionally with fictional events and characters depends on the life concerns we carry in the real world.[15] As Jenefer Robinson makes this point in her discussion of emotional involvement with characters in literature: 'it is clear that I won't experience any emotional response to a novel unless I sense that my own interests, goals, and wants are somehow at stake' (2005: 114). That is, we come to feel characters' fears, agonies and alienations because we perceive (consciously or subconsciously) the portrayed situation relevant for our own well-being. Our psychological affinity with fictional characters allows us to feel what they feel.[16] Richard and Bernice Lazarus succinctly make this point: 'If their plight were not like our own, we would not react' (1996: 131). We cannot emote with fictional characters unless we sense that our own personal concerns are also at stake. For example, not all of us are able to emote with the main character of the movie Into the Wild (2007), which tells a story of a Christopher McCandless (portrayed by Emile Hirsch), who decides to abandon all his possessions and travel to Alaska to live in solitude with the wild nature. One who has never felt the intense longing for freedom, for self-discovery, and for a detachment from humanity may find it hard to imagine McCandless's state of mind and to understand his behaviour. Our ability to mentally simulate psychological states of characters depends on our recognition of psychological similarity with them. This shows that narrative art directly taps on our own emotional and psychological system. How we come to respond emotionally to characters and events portrayed in narrative works represents a unique mode of access to the entire domain of our own personal life concerns, which are often left unacknowledged by us. For example, it is often the case that we feel various emotions, say, fear or hopelessness, yet without being able to articulate fully and understand what it is about the situation that we experience as threatening or hopeless. That is, we can have introspective awareness of our own internal experience yet without being fully aware of the various psychological aspects that comprise the background of such experience. Narrative art, however, can make these psychological aspects salient, thereby available for our acknowledgement.

As pointed out previously, narrative art functions as a simulation of characters' experiences and actions in a compressed format. That is to say, narrative art presents the chain of events in a continuous and complete manner, making salient the relationship between various psychological aspects that comprise the background of characters' emotional experience (that is, how is the depicted situation relevant to the characters' purposes, beliefs, desires and aspirations) and thereby renders those experiences and actions more comprehensible to us.[17] The formal and stylistic properties of narrative art give us the opportunity to observe characters' experiences from a third-personal perspective and help us grasp the meaning of their experiences in the broader context of their mental life. Now, given our own emotional and psychological affinity with fictional characters (without which we would not be able to mentally simulate characters experiences), this suggests that narrative art can also help render our own emotional experiences more intelligible to us. In other words, if our own emotional appraisal of the depicted situation fits in some respect with characters' appraisals (that is, we experience the same or similar emotional reactions to the depicted events as characters do), then narrative art can, by offering the interpretation of characters' experiences, also help us clarify our own emotional experiences. Narrative art can thus give us the opportunity to obtain third-personal (spectatorial) perspective on our mental states, by means of which we can acknowledge and examine certain aspects of our experiences that are left unnoticed from a mere phenomenological first-person perspective. As H. W. Johnstone states, to become aware of certain aspects of our own mental state a certain distance must be inserted between us and our mental state, because '[i]n complete immersion in experience there is no sense of ownership' (1970: 106). Self-knowledge requires a dialogical relationship – the interplay between 'the familiar and the strange, the old and the new, the self and the non-self' (Felski 2008: 49). It is precisely the artistic representation of feelings, state of affairs, ideas, values and beliefs that opens such a dialogue between us, our own subjective experiences (say, how the feeling of emotional isolation is felt by me) and a particular interpretation of such experience (say, how Haneke's film presents the idea of emotional isolation). A dialogue enhances a distance between our subjective state and the objective vision of that mental state through which our own perspective can be revealed. Such mental distance can give us the opportunity to acknowledge and examine certain aspects of our experience that are left unnoticed from a mere first-person perspective. Just like seeing our own body image in the mirror can give us information about our body that cannot be obtained by mere phenomenological experience of our body, so too viewing our own subjective experience from a third-person point of view can give us information about our own experience that is impossible to obtain by first-personal perspective.

I wish to emphasize here again that narrative artworks as mental simulations promote self-knowledge not because they mirror our own lives, but rather because of the way or manner that characters and their situational aspects are presented. This is a distinctive aesthetic quality for it requires artist's choice of particular formal features through which the content is presented. Accordingly, works of art promote self-knowledge precisely because they are substantially different from real-life narratives. They are opaque, rather than transparent, as Peter Lamarque writes:

> Rather than supposing that narrative descriptions are a window through which an independently existing (fictional) world is observed, with the implication that the very same world might be presented (and thus observed) in other ways,

from different perspectives, we must accept that there is no such transparent glass – only an opaque glass, painted, as it were, with figures seen not through it but in it.

(2014: 3)

Artistic narratives represent fictional characters and their situational aspects not transparently, as, for example, photography represents its subject matter; rather, the identity of characters is itself determined by the mode or form of their presentation and which reflects author's point of view or thoughts about the psychological and emotional situation of characters. Each descriptive detail is carefully created, selected and causally related for the purpose of providing thematic coherence through which the author offers their own explanation of the depicted events. It is through the lens of the author's perspective that we come to understand the fictional world described in the work's content: 'Literary works are finely wrought artistic structures affording a special kind of internal connectedness. The consonance of subject and theme [...] is an end in itself in a way that cannot be applicable to narratives that have the further purposes of describing and explaining the lives of real individuals' (Lamarque 2014: 80). We do not ordinarily experience ourselves in a narrative way, that is, thinking of ourselves as characters in the story we tell about ourselves.[18] Even if we do – for example, when we recount certain events in a story-like manner or in moments of self-reflection – the purpose of such real-life narratives is primarily to convey the accuracy of information, rather than to seek the consonance between the events and the theme: 'We do tell stories about ourselves but more often than not they are mundane, fragmented, inconsequential, and for the most part blandly true rather than grandly inventive' (Lamarque 2004: 403). The formal structure given to the sequence of events in artistic narratives serves the purpose of providing coherence and unity between the events of the story and the theme of the work. For this reason, artistic narratives seek causal connection between thoughts, emotions and actions in order to provide a kind of synthesis of otherwise seemingly unrelated elements of the human world.[19] As David Carr nicely captures this idea, rather than describing the human world, an artistic narrative 'redescribes it' (1986: 120). It is precisely such redescription of the human world, in the sense of making salient the causal, explanatory and inferential relationships between various mental aspects, that allow works of art to afford us with the opportunity to obtain self-knowledge. After all, as Martha Nussbaum argues, our emotional experiences have a narrative structure and personal history, and thus to understand our own emotional responses we must take this structure into account – the relationship between our own personal beliefs about the world, personal life-values and emotions (2001: 236). By means of employing rich narrative and formal structure, such as making salient the temporal and causal connection between emotions and their underlying thoughts and personal life-concerns, works of art provide us with information about the history of one's emotion that we could not otherwise obtain. Narrative art offers different explorations of one's emotional histories in light of which we can understand and comprehend our own emotional experiences. 'When we grasp the patterns of salience offered by the work, we are also grasping our own possibilities' (2001: 243). It invites us to 'try' on various interpretative perspectives on a particular emotional experience and 'test' their emotional resonance with our own emotional histories:

For example, Haneke's movie offers a particular interpretation of subjective experience of emotional isolation and alienation, namely, as the feeling of being imprisoned by the life of daily routines. By specifying the feeling of emotional isolation as one of imprisonment by daily routines, the film provides us with a more refined cognitive vocabulary by means of which we can grasp and comprehend the inner experience of the characters. Yet, in the case that we succeed in imagining characters' emotional experience of alienation as our own, this suggests that the film also facilitates third-personal perspective on our own mental states, thereby allowing us to evaluate the meaning of our feelings of emotional isolation in the larger context of our life and to see more clearly the connection between our particular emotional experience and other mental aspects.

The third-personally obtained information we acquire from the work, however, does not remain unchecked from our first-personal perspective. We do not blindly accept the interpretation offered by the work; we 'try' it out and test how well it agrees or fits with our own experience. Testing the appropriateness of the interpretation occurs through the process of mental simulation itself. As we mentally simulate characters' mental states and experiences, we also simultaneously test the simulated feeling against our own feeling of emotional isolation. This process is not intentional; it is something that occurs spontaneously throughout the mental simulation. When we mentally simulate characters' experiences, we do not lose our own separate sense of self, but still maintain the distance between our own personal identity and characters' psychological aspects. Amy Coplan describes this aspect of mental simulation as the 'self-other differentiation', which 'allows the reader to simultaneously simulate the character's psychological states and experience her own separate psychological states' (2004: 148). Mental simulation allows us to test the simulated mental states and experiences against our own personal characteristics and to sense (feel) the appropriateness or inappropriateness of the given interpretation (say, of experience of emotional isolation) for our own felt experience (say, how we personally experience emotional isolation). If the given interpretation emotionally resonates with our own experience in the sense that makes it appear more clear, determinate and intelligible, then we have a good reason to believe that this interpretation accurately describes our own experience. In other words, we exercise affective first-person authority over the information that we acquire from the narrative work. It is our own feeling of appropriateness that serves as a criterion of the correctness of the information. The decision as to whether the given interpretation fits with our own psychological state is up to us and our own feeling of appropriateness. Such affectively avowed interpretation, as pointed out previously, will also have a modifying effect on the nature of our experience itself.

Using the example of Haneke's movie, if during mental simulation we come to see or feel that the interpretation of the emotional isolation as the feeling of being imprisoned by the life of daily routines more appropriately describes our own feeling of emotional isolation, then this interpretation will, in fact, change the way we feel. The identification of our own experience of emotional isolation as one of imprisonment by daily routines will necessitate restructuring of our own experience so that it fits with the imprisonment principle, that is, with the newly identified belief that what we are feeling is imprisonment. While previously what we felt was some vague sense of emotional isolation, insignificance and detachment from the world, now, after accepting the new interpretation, we come to experience more acutely and sharply a feeling of confinement and a sense of being restrained by our daily routines. Our own feeling of emotional isolation changes (becomes more refined and specified) by accepting the new and more refined interpretation.

Furthermore, once we come to identify our own feeling of emotion isolation in a more concrete and situation-related sense of imprisonment by daily routines, we gain the opportunity of obtaining a further insight into other aspects of our experience, as well as a better understanding of the situational aspects we are up against. As Charles Taylor nicely points out, what we experience is never separated from our won being in the situation:

> Experiencing a given emotion involves experiencing our situation as being of a certain kind or having a certain property. But this property cannot be neutral, cannot be something to which we are indifferent, or else we would not be moved.

> Rather, experiencing an emotion is to be aware of our situation as humiliating, or shameful, or outrageous, or dismaying, or exhilarating, or wonderful; and so on.
>
> (1985: 48–9)

We come to experience the situation in a certain way, say, as humiliating or shameful because of our particular psychological make-up (desires, goals, motivations). It is the articulation of an experience that reveals these complex concrete person-situation intricacies. As the artistic expression provides us with a more adequate representation of our own psychological situation, it helps us uncover aspects of our personality that have been discounted or left unacknowledged – what we truly desire, what motivates us and what we value. The sense of imprisonment by daily routines can be felt only by those of us with a need for agency and control in our own life. Recognition of our true beliefs, feelings, desires, interests and goals opens up new perspectives on our own situational and interpretative aspects and, thus, new behaviour possibilities. Once we come to understand our own mental states, we are no longer in a position to think about ourselves as passive and powerless carriers of those mental states. This is the value of self-knowledge as specifically pointed out by Stuart Hampshire:

> The range of the emotions, feelings and attitudes of mind, identified and distinguished from each other, changes as the forms of human knowledge change. We identify new emotions and attitudes that have never been recognized before. With a new self-consciousness, and with the extended vocabulary that goes with it, we discover new motives for action and new objects to which practical intentions are directed.
>
> (1982: 244)

Self-knowledge grants us an awareness of seeing ourselves as authors of our own mental states and, as such, possessing the ability to control and modify these states. For example, we can come to see our daily routines not as something that we are imprisoned by, but rather as something that we can choose to build our lives around. First-personally accounted and reflected mental states alter the nature of these mental states and have a liberating effect on our personality and behaviour, as we can now recognize various possibilities for our actions.

It is important to point out, however, that not all narrative works of art serve as a source of therapeutic self-knowledge, as not all of them give us the opportunity to adopt the first-person perspective on characters' experiences. As I mentioned previously, each artistic expression of an aesthetic idea offers a different organization of an experience and thus a different meaning of an experience. This means that not every artistic expression will be recognized as an explication of our own inner experience and thereby lead to a potentially beneficial self-information. The failure to engage in the imaginative process of simulating characters' psychological and emotional states is often due to the failure of our own imagination. Mental simulation, as pointed out, is facilitated by the perception of emotional and psychological similarities with fictional characters. Our perception of dissimilarities with fictional characters, on the other hand, hinders this process. It is difficult to imagine the point of view of characters whose mental states and experiences differentiate from ours. We often fail to activate mental simulation when confronted with works that depict horrifying and traumatic experiences. The failure of imagination is due to our unwillingness to simulate thoughts and experiences that are just too difficult for us to handle and that go beyond the realm of our own personal experiences. Sometimes we fail to mentally simulate the depicted events not because these events are too remote from our own personal experiences, but because they are too familiar to us. As Rachel Cooper points out, we have problems simulating traumatic events that remind us of our own traumatic experiences, as such simulation is simply too upsetting and unpleasant for us (2014: 75). For example, someone who has been traumatized by war might have difficulties simulating the horrors of combat as depicted in Oliver Stone's film Platoon (1986). But if we are unable to activate mental simulation and thus to experience the first-person perspective on fictional characters, then we also fail in obtaining first-personal perspective on our own mental states. In other words, we fail to exercise affective first-person authority, to assess the appropriateness or inappropriateness of the given interpretation for our own experience, which is necessary for obtaining therapeutic self-knowledge.

There is yet another scenario in which we fail to obtain therapeutic self-knowledge in spite of our successful mental simulation of characters' experiences. This happens when the simulated experience fails to accord with our own experience. Consider, for example, Lars Von Trier's film Antichrist (2009), which tells a story of a couple who retreat to a remote cabin in the woods in order to heal from the recent death of their child. The story of a loss and grief is familiar to almost all of us, and we have no difficulties imagining the depicted events from the point of view of female character as she experiences deep psychological suffering and paralysing feelings of grief and guilt. Yet, as the story progresses and reveals the meaning of her experience of grief as originating in her beliefs about female nature as inherently evil and malevolent, we find it more difficult to relate personally to her experience. An interpretation of grief as connected with ideas of evil and misogyny appears to be too alien from our own experience of grief. Thus, simulated experience of grief fails to resonate emotionally with our own mental states and experience. We come to see or feel that the interpretation does not fit with our own experience of grief as it brings less sense and intelligibility to our own state of mind. Here we have an example of an affectively disavowed interpretation. Even though such an interpretation is the result of our successful exercise of the affective first-person authority, it does not produce any modification of our self-concepts but merely eliminates the sense of experience that does not hold for us. Thus, while affectively disavowed interpretation does provide a potentially beneficial self-information (as to what our experience is not), it does not promote self-change and thus does not serve as a source of therapeutic self-knowledge.

To conclude, my aim in this chapter was to argue that works of art, as expressions of an aesthetic ideas, can promote the kind of self-knowledge that leads to self-development, that is, therapeutic self-knowledge. I argued that narrative art gives us a unique opportunity to adopt a dual perspective on the self, thereby allowing us to (i) fully acknowledge our own emotional experience from a first-person point of view and (ii) recognize the meaning of our experience as it figures in relation to our own personal characteristics and life concerns (third-person point of view). Accordingly, narrative art enhances our self-exploration by giving us the opportunity to turn inward and reflect on the content of our own subjective experiences. It engages us in a cognitive process of identifying our own personal characteristics, challenging our emotional, social and intellectual patterns and recognizing inadequacies in the thoughts we attribute to our lives and experiences of ourselves and others.

4

Cognitive value of representational and non-representational art

I argued thus far that works of art defined in the Kantian sense as expressions of aesthetic ideas have a cognitive value as they contribute to better cognitive processing and (objectual) understanding of abstract phenomena, as well as helping us to obtain the kind of self-knowledge that leads to self-development (that is, therapeutic self-knowledge). My aim in the present chapter is to consider the relationship between particular forms of art, such as literary, visual and musical art, and their respective cognitive effects. This discussion will rely on Kant's division of beautiful arts in terms of their ability to express aesthetic ideas. While all works of art must express aesthetic ideas in order to have cognitive value, not all works of art have the same expressive and thus cognitive effects in terms of promoting objectual understanding. During this discussion certain discrepancies within Kant's text will also be revealed and answered, such as his apparently inconsistent views on the relationship between non-representational art and aesthetic ideas. In order to solve these discrepancies, I will propose a distinction between productive and reproductive aesthetic ideas communicated by artistic form and mere sensations (mere tones and colours) respectively.

Kant's hierarchy of beautiful art

In §51 Kant offers his classification of beautiful (fine) arts in terms of their ability to express aesthetic ideas. Based on the analogy between art and 'the kind of expression that people use in speaking in order to communicate to each other', he divides all forms of beautiful arts into three different kinds of artistic expressions – expression by means of words, gesture and tone (KU 5:320; 198). In this division of beautiful arts, Kant ranks the art of speech (or literary art) the highest for it 'expands the mind by setting the imagination free and presenting, within the limits of a given concept and among the unbounded manifold of forms possibly agreeing with it, the one that connects its presentation with a fullness of thought to which no linguistic expression is fully adequate, and thus elevates itself aesthetically to the level of ideas' (KU 5:326; 203–4). Pictorial art, which includes paintings and plastic art (sculpture and architecture), ranks the second as they express aesthetic ideas through 'sensible intuition' rather than 'through representations of the mere imagination, which are evoked through words' (KU 5:322; 199). Both paintings and plastic art are similar in that they 'make shapes in space into expressions of ideas' (KU 5:322; 199), yet, whereas plastic art depends on both senses of sight and touch, painting relies on the sense of sight alone. Even architecture, included in the category of plastic art, must express aesthetic ideas in order to count as beautiful art. Architecture, however, expresses aesthetic ideas to a lesser degree than sculpture, because 'a certain use of the artistic object is the main thing, to which, as a condition, the aesthetic ideas are restricted' whereas in sculpture 'the mere expression of aesthetic ideas is the chief aim' (KU 5:322; 199). Finally, Kant attributes the lowest expressive and cognitive value to the art of the beautiful play of sensations, which include absolute music as the beautiful play of sensations of hearing and art of colours (or visual abstract art) as the beautiful play of sensations of sight. While he ascribes these art forms the highest rank when considered as merely agreeable art and thus as providing mere enjoyment to the senses of sight and hearing, they nevertheless occupy the lowest rank when considered as beautiful art and as expressive of aesthetic ideas.[1] As he writes in respect to music:

> If, on the contrary, one estimates the value of the beautiful arts in terms of the culture that they provide for the mind and takes as one's standard the enlargement of the faculties that must join together in the power of judgment for the sake of cognition, then to that extent music occupies the lowest place among the beautiful arts [...] because it merely plays with sensations.
>
> (KU 5:329; 206)

Accordingly, not all forms of art express aesthetic ideas to the same degree and thus not all of them can produce the same cognitive effects. Non-representational art, such as absolute music (music without words) and abstract visual art, occupies the lowest cognitive status, presumably, because they are mere beautiful play of sensations.

Unfortunately, Kant does not offer any explanation of the connection he makes between the beautiful play of sensations and its low ability to express aesthetic ideas. Moreover, this connection appears to be highly confusing when considered in light of Kant's claim that it is the form, rather than the content of the object that functions as the 'vehicle of communication' of aesthetic ideas (KU 5:313; 191). As he states explicitly, 'in all the beautiful art what is essential consists in the form, which is purposive for observation and judging, where the pleasure is at the same time culture and disposes the spirit to ideas' (KU 5:326; 203). Yet, Kant's notion of the beautiful play of sensations (in contrast to mere agreeable play of sensations) does refer to the synthesized play of sensations in time (absolute music) and space (abstract art), and accordingly to the mere form of the object. Thus, there is prima facie no reason to assume that non-representational art, which consists in the mere form of the object, that is, in the temporal and spatial arrangements of tones, colours, shapes, lines without any representational content, should express aesthetic ideas to a low degree.

Some have argued that non-representational art expresses low degree of aesthetic ideas because it belongs to the category of free beauty, in contrast to adherent beauty. In short, free beauty refers to the beauty of an object in virtue of its form alone without presupposing what the object ought to be, whereas adherent beauty takes into account the concept of a purpose and therefore 'the perfection of the object in accordance with it' (KU 5:229; 114). As Kant writes in §16, 'designs a` la grecque, foliage for borders or on wallpaper, etc., signify nothing by themselves: they do not represent anything, no object under a determinate concept, and are free beauties. One can also count as belonging to the same kind what are called in music fantasias (without a theme), indeed all music without a text' (KU 5:229; 114). Based on this, Martin Weatherston concludes that representational forms of art, which belong to the category of adherent beauty as they presuppose the concept of what the object ought to be or represent, have a higher status in terms of expressing aesthetic ideas: 'While a pure and free judgement of taste merely assesses the harmony of the imagination and the understanding, judgement upon adherent beauty furthers the culture of the mental powers [...] From this we can see that free beauty is ultimately of lower value than adherent beauty, since it is connected to no concept' (1996: 58).

However, such an interpretation appears to be unconvincing in light of Kant's view of naturally beautiful objects, which, even though belonging to the category of free beauty as they presuppose no concept of a purpose, they nevertheless express aesthetic ideas to a high degree.[2] As he states: 'In general, the beauties of nature are most compatible with the first aim [disposing the spirit to ideas] if one has become accustomed early to observing, judging, and admiring them' (KU 5:326; 203). Similarly, he writes in §42:

> If a man who has enough taste to judge about products of beautiful art with the greatest correctness and refinement gladly leaves the room in which are to be found those beauties that sustain vanity and at best social joys and turns to the beautiful in nature, in order as it were to find here an ecstasy for his spirit in a line of thought that he can never fully develop.
>
> (KU 5:229; 179)

The value of natural beauty in terms of disposing the spirit to ideas far surpasses the value of artistic beauty, even though, as Kant adds, artistic forms can be more beautiful than natural forms: 'This preeminence of the beauty of nature over the beauty of art [...] even if the former were to be surpassed by the latter in respect of form' (KU 5:229; 179). Accordingly, the low degree of aesthetic ideas expressed by non-representational art cannot be attributed to their lack of representational content and the concept of the purpose. Besides, in §48 Kant claims that all artistic beauty presupposes the concept of a purpose and therefore is of adherent kind. Given that non-representational art is made with a certain purpose in artist's mind, namely, to exercise free imagination in the play of colours, tones and forms, its beauty must also be of adherent kind, even though it does not include any representational content.[3] There is therefore no correlation between free/adherent beauty distinction and low/high-degree expression of aesthetic ideas.

A different attempt to explain the relationship between non-representational art and its low expressive status has been made by Emine Hande Tuna (2019). She argues that in contrast to representational art, which is capable of representing noble moral ideas of reason, non-representational works of art as mere beautiful play of sensations can present only ideas of affects and emotions (such as ideas of sadness, anger, cheerfulness, hope, fear and sorrow). Presumably, the lack of connection between non-representational art and rational ideas is the main reason lying behind Kant's attribution of low expressive and cognitive value to absolute music and art of colours. However, such interpretation does not appear to be supported by Kant's textual evidence, in particular with his statement in §42 that mere sensations such as colour and tones can express a variety of rational and moral ideas:

> The white color of the lily seems to dispose the mind to ideas of innocence, and the seven colors, in their order from red to violet, to the ideas 1) of sublimity, 2) of audacity, 3) of candor, 4) of friendliness, 5) of modesty, 6) of steadfastness, and 7) of tenderness. The song of the bird proclaims joyfulness and contentment with its existence.
>
> (KU 5:302; 181)

In fact, Kant's claim that mere sensations can express aesthetic ideas, as well as his highly ambiguous view on the expressive and cognitive status of non-representational art, points out two further inconsistencies related to the nature of aesthetic ideas themselves. First, on the one hand, he appears to be committed to the view that mere sensations, such as a mere tone or a colour, can express aesthetic ideas. For example, as he writes in the passage given above, 'the white color of the lily seems to dispose the mind to ideas of innocence' (KU 5:302; 181). Elsewhere he also states that 'sensations, each of which has its relation to affect ... arouses aesthetic ideas' (KU 5:331; 208). Or, that a mere tone 'more or less designates an affect of the speaker and conversely also produces one in the hearer, which then in turn arouses in the latter the idea' (KU 5:328; 205). Yet, on the other hand, he seems to suggest the contrary when he claims that 'the matter of the sensation ... leaves behind it nothing in the idea, and makes the spirit dull' (KU 5:326; 203). One might attempt to resolve the inconsistency regarding the ability of mere tones and colours to express aesthetic ideas by referring to §14 where Kant discusses Euler's theory and opens the possibility that tones and colours are not mere sensations, but rather play of sensations (form of intuition) (KU 5:224; 109). If tones and colours involve the form of intuition and thus require the activity of imagination, then there is no inconsistency in claiming that they can express aesthetic ideas. However, besides the fact that Kant never fully affirms Euler's theory, he also argues in §51 that even though tones and colours concern form of intuition, we nevertheless perceive them as mere sensations, because 'the rapidity of the vibrations ... far exceeds all our capacity for judging immediately in perception the proportion of the division of time' (KU 5:324; 202). That is to say, the speed of the vibrations is too fast for us to perceive and thus what we end up sensing is merely the 'effect of these vibrations on the elastic parts of our body' (KU 5:325; 202) rather than their temporal form. Thus, the inconsistency remains.

Second, he writes that non-representational art, such as absolute music (in other words, composition of sensations in time), is the kind of 'play with aesthetic ideas or even representations of the understanding, by which in the end nothing is thought' (KU 5:332; 208) and 'does not, like poetry, leave behind something for reflection' (KU 5:328; 205). However, the possibility of an object expressing aesthetic ideas without leaving behind any thoughts to reflect upon appears to be inconsistent with Kant's initial definition of an aesthetic idea as a 'representation of the imagination' that 'by itself stimulates so much thinking that it can never be grasped in a determinate concept, hence which aesthetically enlarges the concept itself in an unbounded way' (KU 5:315; 193). The distinctiveness of aesthetic ideas lies precisely in their ability to generate such richness of thoughts and meanings that yields (aesthetic) expansion of our concept – that is, new ways of understanding and interpreting them. On this view, aesthetic ideas are essentially reflective imaginative representations. They are, as Kirk Pillow puts it, 'the outcome of an expansive reflection' (2001: 200) on a particular presentation and they come to express 'a new perspective that transforms our understanding of ourselves and our world' (2001: 202). Accordingly, how can we reconcile these two inconsistencies within Kant's text?

Unfortunately, the interpretations given so far have failed to provide an answer to this question. While much ink has been spilled, and rightly so, over explaining Kant's ambiguous notion of music as being both agreeable and beautiful (Schueller 1955, Kivy 1993, Weatherston 1996, Parret 1998, Davidson 2000, Matherne 2014, Tuna 2019, Young 2020),[4] the discrepancies within Kant's theory of aesthetic ideas itself have been left unaddressed.

My aim in what follows is to reconcile these inconsistencies by proposing a distinction between the two kinds of aesthetic ideas, namely productive and reproductive aesthetic ideas. I believe that when Kant introduces aesthetic ideas as representations of imagination that stimulate much thoughts, he is referring to productive notion of aesthetic ideas, which are product of the free productive imagination. It is distinctive about them that they not merely give rise to variety of ideas that go beyond sensory experience but since they are carried forward by the freely imaginative and thus original associational thoughts (that is, aesthetic attributes), they also provoke further reflection on these ideas. Productive aesthetic ideas thus have a cognitive value in that they contribute to the expansion of our cognitive faculties (KU 5:329; 206). Aesthetic ideas enlarge our faculties in the sense that they provide (by means of aesthetic attributes) a sensible counterpart to ideas that go beyond our sensory experience (say, an image of an eagle being an indirect sensible counterpart to the rational idea of the king of heaven), or they connect the introspective, emotional and affective properties associated with our abstract concepts with concrete and imaginable representations and thereby provide an additional source of information that is required for a more complete understanding of abstract phenomena.

Yet, there is also a more primitive, non-reflective sense of aesthetic ideas. An object can point to something beyond itself by simply resembling some other aspects from the human world. For example, slow, dark and heavy movements of tones can bring to mind the idea of sadness or tragedy by means of resembling the tonal and bodily expression we make when we feel sad. Or, the blue colour and texture of Yves Klein's painting *La Vague* (1957) can evoke the idea of limitlessness or infinity because it resembles the colour and texture of the wide sea. Sensations, such as tones, colours and textures, can bring to mind a variety of ideas that go beyond sensory experience, but only by means of associations with which we are already acquainted. They merely reproduce or recall already-existing associations, rather than offer an imaginatively rich and original semantic material that would further our reflection on these ideas themselves. I believe such an interpretation is implicit in Kant's claim that 'the matter of the sensation ... leaves behind it nothing in the idea, and makes the spirit dull' (KU 5:326; 203). His use of the words 'leaves behind it nothing in the idea' suggests that there must be an aesthetic idea expressed by sensations in the first place; yet, no afterthoughts about these ideas are left behind. Hence, sensations can express merely reproductive aesthetic ideas that fail to produce objectual understanding.

This is the thesis that I will argue for in what follows. In particular, I will attempt to show that (i) both mere sensations and form of the object can express aesthetic ideas, but it is only the latter that can express productive aesthetic ideas that have an inherent cognitive value; (ii) productive aesthetic ideas can be exercised to different degrees depending on the degree of the free (productive) imagination incited by the form of the object; and (iii) in comparison to representational art, non-representational art, such as absolute music and abstract visual art, allows for a low degree of free imagination and thus expresses productive aesthetic ideas to a low degree. Consequently, non-representational art is of lesser cognitive value than representational art.

Sensations and reproductive aesthetic ideas

We know so far that aesthetic ideas are a product of the synthesis of imagination, whereby the components that comprise this synthesis are aesthetic attributes – in other words, thoughts, associations and other mental representations evoked by the features of an object. This implies that form of the object is essential for the communication of aesthetic ideas.

However, in his discussion of music Kant introduces the idea that mere sensations can also express aesthetic ideas. As he writes:

> every expression of language has, in context, a tone that is appropriate to its sense; that this tone more or less designates an affect of the speaker and conversely also produces one in the hearer, which then in turn arouses in the latter the idea that is expressed in the language by means of such a tone; and that, just as modulation is as it were a language of sensations universally comprehensible to every human being, the art of tone puts that language into practice for itself

alone, in all its force, namely as a language of the affects, and so, in accordance with the law of association, universally communicates the aesthetic ideas that are naturally combined with it.

(KU 5:328; 205)

In this passage Kant makes three points. First, each musical tone, on the account of being mere sensation, expresses a particular affect in much the same way as the tone of the speaker expresses his or her emotional state (for example, loud, bright and energetic tone of voice expresses the feeling of cheerfulness or joy). Second, the affective component expressed by a specific musical tone in turn arouses in the hearer a specific affect-related aesthetic idea (for example, the idea of cheerfulness). Third, these affect-related aesthetic ideas are a product of the empirical law of association or, as Kant adds in the same passage, they are 'merely the effect of an as it were mechanical association' (KU 5:328; 205). Presumably, the association between musical tones and affect-related aesthetic ideas is mechanical, that is, automatic and spontaneous, because it is a product of mere resemblance or similarity between the structure of musical tones and tonal structure of language and its expressive affective qualities.[5] For example, loud and energetic tone stimulates the idea of cheerfulness because it simply resembles the sounds we make when we feel cheerful. The law of association, in particular, the law of similarity or resemblance, simply connects past experience (in other words, remembering the sound of a cheerful person) with our present experience of the musical tone. Accordingly, the suggestion seems to be that while mere sensations, such as musical tones, can express aesthetic ideas in the sense that they bring to mind ideas and concepts that go beyond sensory experience (for example, the idea of cheerfulness or joy), they do so merely by means of reproducing or bringing to mind already-existing associations, rather than by presenting original associational thoughts (in other words, aesthetic attributes).[6] Thus, while musical tones can express well the phenomenology of a particular affect-related idea (say, how cheerfulness or sadness feels like subjectively), they lack the ability to further our reflection on these ideas themselves. The kind of aesthetic ideas that musical tones can express are, as I call them, reproductive aesthetic ideas.

The case is similar with mere colours, which Kant treats analogously to musical tones. Colours tend to express aesthetic ideas by means of habitual associations, being the product of resemblance. For example, the colour red is strongly associated with ideas of vitality, life, power and passion by means of resembling the colour of human flesh and blood, which have life-giving properties. Or, the colour yellow is associated with the ideas of optimism, new beginnings, hope and joy on the account of being a colour of the sun, which is the source of warmth, nourishment, life and growth.

I wish to point out that reproductive aesthetic ideas differ from what Kant calls symbolic exhibition of ideas (in other words, indirect analogical presentation of ideas) even though they both rely on a pre-existent association – namely, association by similarity. According to Kant, symbolic exhibition consists in comparing two apparently dissimilar, yet conceptually analogous things, such as hand mill and despotic state. Even though there is no obvious similarity between the empirical object hand mill and the concept of despotic state, there is however a similarity in the definition of both in that 'the material fed through a hand mill is to its operator as the subjects under an absolute monarchy are to the despot' (Pillow 2001: 194). We can see that symbolic exhibition depends on a pre-existent, yet hidden and non-obvious, similarity between two different things and the symbol-maker must exercise his own productive imagination in discovering this similarity or bringing it into explicit awareness. The case is different with reproductive aesthetic ideas as they depend on an apparent, clearly noticeable similarity between two things (for example, there is an obvious similarity between slow movement of tones and sad bodily expression) with the artist depending solely on the reproductive imagination (in other words, mere series of empirical/mechanical associations) recalling or reproducing it. Unlike symbolic exhibitions, reproductive aesthetic ideas elicit an automatic association between sensations and ideas. Loud and energetic tones automatically trigger the idea of cheerfulness because they directly resemble the sounds we make when we feel cheerful. The association between a particular tone and the affect-related aesthetic idea is immediate and does not require any reflection.

However, absolute music can express not only reproductive, but productive aesthetic ideas as well, considering that it consists not merely of the matter of sensation, but of form as well – in other words, temporal organization of the matter of sensation (KU 5:225; 110). For example, Kant writes that music can be appreciated not merely as an un-synthesized play of sensations (music as agreeable art), but also as a synthesized play of sensations in time governed by a mathematical rule (music as beautiful art) and which gives rise to 'the aesthetic ideas of a coherent whole of an unutterable fullness of thought, corresponding to a certain theme, which constitutes the dominant affect in the piece' (KU 5:329; 206). Presumably, musical form expresses the aesthetic idea of the dominant affect, which refers to the overall theme of the work. Given that music is a combination of tones and each tone or a movement of tones is associated with a particular affective idea, this means that aesthetic idea of the dominant affect refers to a kind of holistic mental representation of various particular affective ideas combined and unified together. Each movement of tones in a musical piece produces certain affect-related ideas in the listener that, when combined together, constitute the dominant affect or emotion.

For example, Bach's musical piece *The Well-Tempereted Clavier I* (1722), consists of twenty-four preludes and fugues, each of them creating its own affective mood that when taken together constitute one complex emotional theme. For instance, the first prelude and fugue begin with C-major, producing the affect of purity, simplicity and joyfulness, being followed by the second prelude and fugue in C-minor, which elicits the mood of danger, until resulting in C-sharp major, which lightly relieves the mood. The affective movement continues with the fourth fugue in C-sharp minor, which in its transparent sadness brings pensive mood and introspection until finally arriving to the affection of gladness, joy and noble dignity of the fifth fugue in D-major. The composition of these particular affects constitutes the dominant aesthetic idea of the work – that is, a concrete presentation of an idea, such as the idea of the internal spiritual and religious struggle. Even though particular affective ideas are the product of mechanical associations (as musical tones are automatically associated with certain affects, such as C-major with happiness or C-minor with sadness) and to this extent they exhibit no free productive imagination, the synthesis of these different particular affective ideas (in other words, the aesthetic idea of the dominant affect) is nonetheless the product of the free play of imagination (as the synthesis is not governed by any determinate concept). As Kant writes, 'it is not the sensation directly (the material of the representation of the object), but rather how the free (productive) power of imagination joints it together through invention, that is, the form, which produces the satisfaction in the object' (Anthro 7:240; 137). That is, even though the work is constituted by particular mechanical associations, this does not necessarily render the combination of these associations' mechanical as well. The latter can nevertheless be the work of a free productive imagination and can thus lead to a genuine productive and cognitively valuable aesthetic idea.

Cognitive value of reproductive and productive aesthetic ideas

Aesthetic ideas are a product of the imagination or, as Kant writes, they are 'only a talent (of the imagination)' (KU 5:314; 193). Yet, while productive aesthetic ideas are a product of the free productive imagination, reproductive aesthetic ideas are a product of the reproductive imagination. Kant explains the distinction between productive and reproductive imagination in the following way:

The power of imagination (facultas imaginandi), as a faculty of intuition without the presence of the object, is either productive, that is, a faculty of the original presentation [Darstellung] of the object (exhibitio originaria), which thus precedes experience; or reproductive, a faculty of the derivative presentation of the object (exhibition derivativa), which brings back to mind an empirical intuition that it had previously.

(Anthro 7:167; 60)

Imagination can produce representations in two different ways. It can either generate original fictitious forms (for example, Picasso creating original imaginative representation of the grieving woman) or it can merely reproduce representations based on past experiences (for example, my imagining Picasso's painting that I previously saw). In the former case, imagination invents images independently from the actuality of objects, whereas in the latter case imagination does not produce new presentation, but merely remembers previously given ones. Reproductive imagination reproduces representations (in other words, calls back previously experienced representations) according to the empirical law of association (V-Anth/Fried 25:512; 81). Kant distinguishes three different ways of associating ideas and objects: accompaniment, contiguity and relation. Accompaniment consists in the repeated succession of representations in time. For example, 'if we see smoke, then the representation of fire immediately appears. If the clock strikes at whichever time one is accustomed to eat, and one hears it striking, then the representation of food immediately appears' (V-Anth/Fried 25:512; 81). Contiguity consists in the association of representations based on their proximity in space: 'If one thinks about the place where one enjoyed oneself, the people who were present there come to mind. If one travels to the place where many events occurred, one remembers these and the mind is stirred up by [the memory of] such events' (V-Anth/Fried 25:512; 81). Association by relation, on the other hand, refers to representations that are related to each other based on their formal constitution. It is comprised of the relation of derivation and relation of similarity. Derivation involves causal relations, such as 'if it rains and the sun shines, one immediately looks around [to see] if there is not a rainbow' (V-Anth/Fried 25:512; 82), whereas similarity involves common feature shared by different representations 'so that if we think of one thing, the other comes to mind' (V-Anth/Fried 25:512; 82)

Reproductive imagination (the ability to make associations) is conditioned by the 'affinity of appearances' (A122). That is to say, we could not make empirical associations in our minds unless appearances exhibit certain regularity, uniformity and coherence. If our experiences did not exhibit any regularity, but were completely chaotic and random, then we could never make any associations between different objects, that is, reproductive imagination would be useless. The affinity of appearances is the objective ground of associative connections (A121). As Kant writes,

> Suppose that cinnabar were now red, then black, now light, then heavy; or that a human being were changed now into this and then into that animal shape; or that on the longest day of the year the land were covered now with fruit, then with ice and snow. In that case my empirical imagination could not even get the opportunity, when presenting red color, to come to think of heavy cinnabar. Nor could an empirical synthesis of reproduction take place if a certain word were assigned now to this and then to that thing, or if the same thing were called now by this and then by another name, without any of this being governed by a certain rule to which appearances by themselves are already subject.
>
> (A101)

As argued previously, reproductive aesthetic ideas are product of the reproductive imagination governed by the empirical law of association, such as the law of similarity or resemblance. Slow and heavy movement of tones is associated with the idea of sadness by means of similarity with the sound we make when we feel sad. Hearing a slow tone simply recalls to our memory the sound of a sad person. Similarly, we associate the red colour in Mark Rothko's painting entitled *Orange, Red and Red* (1962) with ideas of life, vitality and energy because of its similarity to the colour of objects that have life-giving properties (for example, red flesh and blood). We can see that reproductive imagination is directed towards what we perceive through the senses and with which it must be already acquainted in order to make associations between different representations. It is a passive process as it merely mechanically reproduces (in other words, recalls or brings to mind) previously given empirical intuitions.

Productive imagination, on the other hand, creates new presentations based on the material given by empirical nature:

> The imagination (as a productive cognitive faculty) is, namely, very powerful in creating, as it were, another nature, out of the material which the real one gives it [...] in this we feel our freedom from the law of association (which applies to the empirical use of that faculty), in accordance with which material can certainly be lent to us by nature, but the latter can be transformed by us into something entirely different, namely into that which steps beyond nature.
>
> (KU 5:314; 192)

Productive imagination is free to the extent that is not determined by the laws of associations which, in other words, means that imagination is 'taken not as reproductive, as subjected to the laws of association, but as productive and self-active (as the authoress of voluntary forms of possible intuitions)' (KU 5:240; 124). The freedom of the productive imagination is measured in terms of the degree to which it is restricted by empirical conditions and its laws of associations. The less restricted by empirical, external objects, the freer the imagination is. As Kant points out this idea explicitly:

> The imagination is more intense there where other senses are weaker. For example, the imagination of a blind person is by far more intense than someone else's, since it is not disturbed by external objects ... The imagination is greater when the object is absent, than when it is present. If the object is present, then the sensible impression is present, and the imagination is obscured by the sensible impression. If however the object is absent, then the imagination is more intense.
>
> (V-Anth/Fried 25:514; 82)

Accordingly, the imagination is weakened by the presence of sense impressions, whereas the removal of external objects heightens the freedom of the imagination.

Within this framework, we can make sense of Kant's classification of beautiful art in terms of their ability to promote the enlargement of our cognitive faculties – in other words, express productive aesthetic ideas. For example, he writes that literary art, such as poetry, ranks the highest, because it 'expands the mind by setting the imagination free' (KU 5:326; 203–204), whereas music claims the lowest rank 'because it merely plays with sensations' (KU 5:229; 206). If productive aesthetic ideas are product of the free productive imagination, whereas the freedom of imagination is measured in terms of the degree to which it is restricted to empirical conditions, it follows that the less imagination is restrained by empirical conditions, the higher is the degree of productive aesthetic ideas. The distance from the sensuousness and the immateriality of art is aligned with the freedom of the productive imagination and with the expression of productive aesthetic ideas.

In this respect literary art ranks the highest – it admits the highest degree of free (productive) imagination, because it is not tied down for its realization to empirical intuition. Literary art communicates aesthetic ideas not through sensible presentations, but through linguistic elements (words and concepts), which allows imagination to be free from the empirical nature (whereby the imagination would be merely reproductive). As Kant states in respect to poetic art 'strengthens the mind by letting it feel its capacity to consider and judge of nature, as appearance, freely, self-actively, and independently of determination by nature' (KU 5:326; 204). Liberated from the external, empirical conditions, the imagination can play freely and spontaneously with the given semantic material, generating original, unexpected and creative combinations of existing concepts and thoughts and thereby elevating to the presentation of ideas. As Kant writes, it is poetic art 'in which the faculty of aesthetic ideas can reveal itself in its full measure' (KU 5:314; 193), the reason being that it 'owes its origin almost entirely to genius' (KU 5:3261 203), that is, to the free productive imagination: '[T]he proper field for genius is that of the power of imagination' (Anthro, 7:224; 120). Due to the exclusion of the appeal of the senses, literary art is essentially the work of pure imagination: 'It declares that it will conduct a merely entertaining play with the imagination, and indeed concerning form, in concord with the laws of the understanding, and does not demand that the understanding be deceived and embroiled through sensible presentation' (KU 5:327; 205).

The case is different with visual art, which admits lesser degree of free imagination because it communicates aesthetic ideas through 'sensible intuition' (KU 5:321; 199). Even though the artist is free to create highly imaginative forms, they nevertheless rely on objects given by empirical nature and to this extent the freedom of their imagination is limited. For example, Kant writes that in representing objects that in nature would arouse disgust, which is an aesthetic emotion that is essentially tied to the condition of the senses (particularly senses of taste, smell and touch), the visual artist must appeal to symbolic representations in order not to destroy aesthetic satisfaction:

> The art of sculpture, since in its products art is almost confused with nature, has also excluded the representation of ugly objects from its images, and thus permits, e.g., death (in a beautiful genius) or the spirit of war (in the person of Mars) to be represented through an allegory or attributes that look pleasing, hence only indirectly by means of an interpretation of reason, and not for the aesthetic power of judgment alone.
>
> (KU 5:312; 190–91)

This is different in the case of literary art which can suspend the anti-aesthetic effect of disgust since representation of disgusting objects through words is more distant from the appeal of our senses. Besides, visual art such as painting incorporates the use of colours, which on the account of being mere sensations appeal to our sensible nature and thereby distract the mind from reflection on aesthetic ideas.[7]

There is yet another reason as to why Kant considers literary art as a work of pure imagination. Namely, as pointed out by Alberto J. Coffa, Kant holds a semantic theory according to which the meaning of words is communicated through mental representations. That is to say, in order to grasp the meaning of the word, say a dog, one must form a mental representation of a dog that one has previously seen, which is not the case if the image of the dog is already given and imagination must merely reproduce it. Meanings for Kant are 'inextricably associated with experience' and 'in order to know the meaning of pain, love, rivalry, heroism, and so on, one must undergo certain experiences' (Coffa 1991: 9). To know the meaning of concepts such as love or pain one must mentally imagine it, that is, mentally represent the semantic content of it. This semantic process, albeit more complexly structured, is at work when the words are carefully arranged together in a poem or any other form of literary art. To grasp the literal meaning of a poem, one must first grasp the meaning of each separate word by forming mental representations that each word brings to mind, and second, the imagination must connect together all the mental representation brought to mind by each word into one complex mental representation in virtue of which the literal meaning of the poem can be communicated. As Coffa writes, Kant holds a chemical theory of representation according to which 'representations, like chemical compounds, are usually complexes of elements or "constituents", which may themselves be complex' (Coffa 1991: 9). Each word-mental representation is itself construed of many different mental representations. For example, the word 'sun' is construed from mental representations of light, round, warm and so on. Similarly, the meaning of the entire work is a complex mental representation construed from its particular elements – words as mental representations.

We can see that in literary art the imagination is fully engaged, not only in generating aesthetic attributes stimulated by the mental representations of words, but also in semantically processing (or mentally imagining) the literal meaning of each word. The imagination must first produce mental representations in order to grasp the semantic meaning of each word and then based on these mental representations further engage in producing aesthetic attributes (or imaginative associational thoughts) that yield an aesthetic idea. That is, the imagination must connect semantically processed mental representation with other concepts, thoughts, ideas and feelings. The case is different in visual art. Given that pictorial art depends on already-given empirical intuitions (images), the function of imagination is restricted to its reproductive role, namely, to reproduce the semantic meaning of the image. No productive imagination is required to semantically process the literal meaning of the image by producing mental representation of it.

Absolute music, on the other hand, admits the lowest degree of free productive imagination because it is a mere play of sensations. For Kant, the concept of 'play' is a temporal notion, that is, it refers to temporal arrangements of sensations: 'Variousness in accordance with time is a play, hence music is a play of sensation' (V-Anth/Fried 25:496; 68). Thus, when Kant describes music as a mere play of sensations, he is referring to music as synthesized play of sensations in time and thus as a form of intuition. However, even though music is a product of imagination as it consists in formal arrangement of sensations (rather than mere sensations), it nevertheless admits low degree of productive imagination (and consequently low expressive and cognitive value) presumably because it is a mere play of sensations in time, in contrast to, say, pictorial art which is a play of sensations in time as well as in space. As Kant states, 'hearing is a play of sensation, and sight a play of shape. Variousness in accordance with time is a play, hence music is a play of sensation. Variousness in accordance with space is shape' (V-Anth/Fried 25:496; 68). Considering that he uses the temporal notion of 'play' for the objects of sight, this suggests that he takes our experience of shape representations in pictures and sculptures as being not merely spatial, but temporal as well.[8] According to Kant's view, it is spatiotemporal arrangement of sensations that constitutes the form of the object: 'the form, the combination of the impressions; this is the way in which the object is determined in a space' (V-Anth/Mron 25:1241; 370). Thus, even though in the third *Critique* Kant divides the form of the object into the play of sensations in time (music and art of colours) and into shape or the play of shapes (KU 5:225; 110), it is actually the latter that constitutes the proper form of the object. We can conclude accordingly that for Kant visual art ranks higher than music because it deals with the objects of sight (in other words, shape representations) and requires the subject's imaginative activity of apprehending spatiotemporal relations of material qualities. As he writes, sight is the noblest sense, because 'not only has the widest sphere of perception in space, but also its organ feels least affected (because otherwise it would not be merely sight). Thus, sight comes nearer to being a pure intuition (the immediate representation of the given object, without admixture of noticeable sensation)' (Anthro 7:156; 48).

On the other hand, our experience of music, on account of being a mere play of sensations in time, is determined less by the activity of imagination and more by the working of our sense organs; as we are affected by material qualities (in other words, tones, which cause vibrations in our body): 'The form of an object is not given through hearing, and sounds of speech do not lead to a direct idea of what the form is' (Anthro 7: 155; 47). Absolute music thus bounders with the agreeable; with 'that which pleases the senses in sensation' (KU 5:205; 91) and it is therefore strongly associated with expression of reproductive aesthetic ideas.

Furthermore, if absolute music expresses reproductive aesthetic ideas, specifically the emotion or affect-related aesthetic ideas, by means of resembling human's vocal and bodily expressions of emotions, and given that vocal and bodily expressions of emotions represent only external symptoms of emotions (such as crying being the external symptom of the emotions of sadness), rather than emotion itself (that is, its aboutness or intentionality), then it follows that music is incapable of representing the meaning of emotional and other abstract concepts and its constitutive cognitive aspects. The structure of music, on this account, is logically incapable of communicating the intentionality of emotions, that is, a way of seeing and interpreting the object of emotion (say, that sadness incorporates the thought or belief that something or someone important to us has been lost), which reflects our own personal characteristics and life-concerns, and determines the identity of the emotion itself.[9] Moreover, if absolute music expresses emotions by means of resembling their affective-behaviour components, then its expressive value is limited not merely in depth, but also in the range of emotions that it can express. Namely, not all emotions can be distinguished by means of their bodily and vocal expressions, since not all of them necessarily have a typical bodily or vocal response (examples are long-term emotions, such as regret, guilt or love; one can feel love, without experiencing any bodily changes).[10] But if a certain emotion does not have a corresponding vocal or bodily response, then absolute music is unable to represent it. In addition, certain emotions exhibit similar or identical vocal and bodily expressions, such as shame and embarrassment, which are both typically associated with withdrawal and hiding behaviours. Distinguishing between the two emotions requires grasping their distinctive cognitive components. For example, shame is constituted by the thought or belief that one has been degraded in a way that casts doubt on their sense of self-worth, whereas embarrassment involves a belief that one is in a socially awkward situation, but which is not necessarily degrading. Accordingly, on this account, absolute music can express merely the affective component of emotion and abstract concepts, but not their semantic content that reflects one's own personal life concerns. If so, music is ultimately incapable of promoting (objectual) understanding of such concepts.

There is yet another reason for the low degree of productive aesthetic ideas expressed by music, namely the transitory nature of thoughts necessitated by the transitory nature of musical sensations (KU 5:328; 205). Kant writes that those sensory impressions left by music 'are either entirely extinguished or, if they are involuntarily recalled by the imagination, are burdensome rather than agreeable to us' (KU 5:330; 207). Presumably, musical temporal nature (made of sounds that vibrate and vanish) and its lack of spatiality that would ground persistency of sensations preclude our ability to keep in mind the occurring thoughts. Hence, thoughts produced by music either very quickly disappear once they occur in time or are turned into fantasies – in other words, involuntary products of imagination (Anthro 7:167; 60), that consist of unrestrained, arbitrary and random movement of thoughts and images, received ready-made from the laws of associations. Accordingly, while music can evoke many affect-related ideas and thoughts as it 'moves the mind in more manifold, and … in deeper ways' (KU 5:328; 205), these thoughts either quickly cease to exist in our mind or they are turned into meaningless fantasies.

Further implications: Kant and abstract art

As pointed out thus far, within Kant's text we can distinguish two kinds of aesthetic ideas: productive and reproductive aesthetic ideas communicated by form and mere sensations respectively. Productive aesthetic ideas are composed of the synthesis of aesthetic attributes (in other words, thoughts, emotions, feelings and other mental representations evoked by the features of an object) and are product of the free and thus creative productive imagination. Reproductive aesthetic ideas, on the other hand, are the product of the reproductive imagination merely recalling or bringing to mind already-existing associations. In this respect, they lack the cognitive value that Kant ascribes to genuine (productive) aesthetic ideas and which consist, according to my proposal, in promoting objectual understanding, that is, comprehending various relationships between introspective, emotional and affective properties underlying abstract phenomena.

The distinction between productive and reproductive aesthetic ideas can help us make sense of some of the inconsistencies within Kant's text. One of such inconsistencies lies in his commitment to the view that mere sensations cannot serve as vehicles for the expression of aesthetic ideas, while at the same time affirming that sensations, such as mere tones and colours, can communicate aesthetic ideas. On my proposal, while mere sensations cannot express productive aesthetic ideas, they can nevertheless communicate reproductive aesthetic ideas. That is to say, sensations can bring to mind a variety of ideas that go beyond sensory experience, but only by means of mechanically reproducing (in other words, automatically recalling) already-existing associations. As such, they cannot further our reflection on these ideas themselves, which furthermore explains Kant's claim that music is an expression of aesthetic ideas, by which nothing in the end is thought. Music, on account of being a mere play of sensations in time (rather than also in space as pictorial art), is highly dependent on the working of our sense of hearing (in other words, how we are affected by tones) and less by the synthetic activity of imagination. Thus, music, as a 'language of sheer sensations' (Anthro 7:155; 47), is highly associated with the expression of reproductive aesthetic ideas. For example, Kant associates each musical tone with a particular affect-related aesthetic idea. Considering his definition of the affect as a 'surprise through sensation' that 'makes reflection impossible (it is thoughtless)' (Anthro 7:252; 50), it follows that even though music can express productive aesthetic ideas, the latter nevertheless serve merely as 'objects for affects' (KU 5:332; 209), thereby failing to make reflection possible. Thus, while music can give rise to a multitude of thoughts – that is, it can express 'the aesthetic ideas of a coherent whole of an unutterable fullness of thought' (KU 5:329; 206) – it fails to leave behind any meaningful thoughts to reflect upon. As Kant states elsewhere, 'Music enlivens us, and promotes and makes it easier for us to dwell on our thoughts better, and is thereby a good motion. But we cannot recount anything about the music' (V-Anth/Mron 25:1331; 437).

Furthermore, I have demonstrated that productive aesthetic ideas can be exercised to different degrees depending on the freedom of the productive imagination incited by the object and which is measured in terms of the degree to which it is subjected to empirical conditions. In this respect, non-perceptual art, such as literary art, expresses the highest degree of productive aesthetic ideas, followed by the pictorial and plastic art, and music, the latter expressing the lowest degree of productive aesthetic ideas, because it is highly dependent on our sense of hearing. The sense of hearing, unlike vision, is determined not by the form of the object, but by 'the matter of the representations, namely mere sensation' (KU 5:224; 108). Hearing, as Kant writes elsewhere, is 'pure sensation' (V-Anth/Fried 25:498; 69), whereas sight 'provides no sensation, but it provides the shape' (V-Anth/Fried 25:496; 68).

It is in this respect, I contend, that Kant would rank abstract visual art higher than music, considering that it consists in the spatiotemporal arrangement of sensations, in contrast to music's mere temporal play, and thus permits persistency and durability of thoughts conveyed. Sight, Kant says, is spatial sense: 'Sight presents the shapes of things in space and divides space' (V-Anth/Fried 25:494; 66). While play of shapes, he writes, 'is still sensation through light and through color', it nevertheless 'pertains more to the object. For this reason we are not as affected' (V-Anth/Fried 25:496; 68–9). However, even though our perception of abstract visual art is determined less by the working of our sense organs (as we are affected by the matter of sensations) and more by the synthetic activity of productive imagination (combining together various shapes, colours, figures, lines), it is still amenable to produce productive aesthetic ideas to a lower degree. This is because it consists of unidentifiable and diverse perceptual forms, which stimulate the imagination to rearrange and reconstruct the shapes, colours, lines, etc., of the manifold without settling on a precise referent and thus it is likely to lead to the production of fantasies and personal idiosyncratic associations, rather than rule-connected associational thoughts and shared semantic associations (aesthetic attributes) that are distinctive for productive aesthetic ideas. James Elkins nicely illustrates such perceptual alternations and accompanying fantasies in his experience of Mark Rothko's fourteen black colour-hued paintings displayed in the Rothko Chapel:

> I looked around. Each of the paintings had a few faded marks, splotches, and odd shapes. None was entirely blank. Many reminded me of dark skies or deep water. (Later, someone told me a visitor had said they are like windows into the night.) I got up and walked over to the next painting. [...] It is full of streaks: cascades of paint that look again like rain, and dull passages that reminded me of high, flat cirrus clouds. [...] I sat down and looked at that painting quite a while, just looking up and down, thinking of clouds and quiet, fine rain.
>
> (2004: 7)

The writer illustrates well the play of thoughts triggered by the play of different shades of black colour. Yet, these kinds of thoughts are inventions not of the free productive imagination that gives rise to productive aesthetic ideas, but rather of involuntary power of imagination or 'fantasies with which the mind entertains itself while it is being continuously aroused by the manifold which strikes the eye' (KU 5:243; 127). Kant describes productive imagination as the 'voluntary power of imagination' (V-Anth/Mron 25:1258; 383) for it is always regulated by the faculty of understanding and thus has a cognitive function.[11] While in aesthetic reflection productive imagination is not governed by concepts of the understanding as it is in ordinary cognition, it does, nonetheless, stand in a certain relationship with the faculty of understanding, that, in a freely playful relationship. Without such relationship between imagination and understanding, no aesthetic experience of the beautiful would arise, nor aesthetic ideas and its reflective effects would take place. Productive, voluntary imagination is responsible for all artistic creations. Fantasies, on the other hand, are involuntary representations of imagination for they lack any connection to the faculty of understanding. As uncontrollable and rule-less products of involuntary imagination, they bring forth images and thoughts deprived of any coherent meanings. Kant gives various examples of such fantasies, which bear a similarity to the play of colours in abstract visual art. For instance: 'looking at the changing shapes of a fire in a hearth or of a rippling brook, neither of which are beauties, but both of which carry with them a charm for the imagination, because they sustain its free play' (KU 5:243–244;127). Or, 'tobacco, with the different indefinite shapes of smoke. [...] So, too, the imagination is served by broad vistas, where I cannot think anything definite about the objects and my fantasy can thus swarm as it pleases' (V-Anth/Mron 25:1259; 384). Kant appears to identify such fantasies and the entertainment they produce to the mind with the feeling of pleasure in the agreeable or charming, this pleasure being grounded on the mere matter of representation (sensations). While the agreeable can sustain the attention, the mind however remains passive for it is deprived of the animating play between cognitive powers of imagination and understanding (KU 5:222: 107), thus failing to produce reflective material. Since the play of thoughts produced by fantasies are not regulated by the faculty of understanding, they, as Kant writes in the passage above, 'swarm as it pleases'. Accordingly, the idea seems to be that while abstract visual art can produce aesthetic ideas, that is, it can bring to mind a variety of concepts, ideas and other mental representations that lack empirical counterpart, these representations are ultimately product of the reproductive imagination, which automatically and thus involuntarily associates the play of shapes, lights, colours and gestural marks with personal fantastic thoughts and ideas: 'With fantasy we often play our game, as we intentionally direct it, but it also plays its game with us, as it carries us away involuntarily toward ideas' (V-Anth/Mron 25:1258; 384). Consequently, visual abstract art fails to further any substantive and comprehensive meanings on the evoked ideas themselves. It merely betrays the presence of a certain idea, but it does not and it cannot present and communicate various causal and explanatory relationships between different aspects that lie at the background of such ideas. Consider, for example, Wu Guanzhong's abstract painting *Alienation* (1992) rendered in black ink and watercolour, which brings to mind ideas of estrangement, emptiness and anxiety. The painting succeeds expressing the qualitative character of alienation – how it feels like subjectively – by means of its mechanical and spontaneous association with black colour and swaying brushstrokes, but it fails to reveal the introspective and emotional features, as well as its implications that form the meaning of the idea of alienation itself. In this respect, the painting fails to satisfy our cognitive interest. This differs, for example, from Edvard Munch's painting *Evening on Karl Johan Street* (1892), which, through the portrayal of corpse-white and faceless mass of people with vacuous expressions, aggressively hurrying towards the viewer as if to attack, succeeds communicating not only the qualitative character of the idea of social alienation, but also its underlying psychological and social mechanism (such as the loss of people's oneness and intimate relation to each other for the sake of blindly pursuing economic ends).

Visual abstract art is similar to absolute music in that it depends on the reproductive imagination and its laws of association (in particular the law of resemblance or similarity) for expression of aesthetic ideas. The difference is that in contrast to music, visual abstract art expresses aesthetic ideas not merely by means of resembling the vocal and behaviour expressions of emotions, but by resembling objects in the human world. In this respect, the depth and range of its expressive power can be slightly greater than in absolute music. For example, a simple line, being an essential building block of abstract visual vocabulary, can express ideas of stability, rest and quiet comfort when placed horizontally on the account of resembling person lying down, yet, ideas of dignity, grandeur and proudness when placed vertically due to its resemblance to a posture of a proud person. Diagonal lines, on the other hand, convey a feeling of instability and ideas of anxious impending events as they resemble a position of an approaching fall (consider, for example, the feeling of restlessness and tension evoked by the strong diagonal composition, separating the houses on the left side and the crowd of people hurrying into the foreground in the Edvard Munch's painting *Evening on Karl Johan Street*). Circles, resembling the shape of life-giving elements like the sun and moon, can represent the cyclical nature of all things – creation and destruction, birth and death (think of Wassily Kandinsky's *Squares with Concentric Circles*, 1913) or ideas of unity, safety and community on the account of being closed, restrictive of the external environment. Accordingly, while visual abstract art can have a richer and specific semantic content when compared to absolute music (since sight is better in identifying things and revealing distinctions), it lacks the ability to reveal and specify the various cognitive and emotional components that constitute the meaning of specific ideas.

In sum, non-representational artworks can express aesthetic ideas and serve as a source of cognitive value, for the visual and aural elements are not purely self-referential and can bring to mind a variety of ideas, which, when taken together, can express aesthetic ideas and thus promote understanding of abstract phenomena. Yet, their cognitive effects are exercised to a low degree, for even though they can exhibit freely imaginative productivity in their formal arrangement, the expressive qualities of the elements (i.e. tones, colours, lines, etc.) that constitute such form depend on already-existing mental associations and as such they fail to offer imaginatively rich and original semantic vocabulary that would provoke further reflection on the expressed ideas themselves. Thus, non-representational art merely reveals the presence of ideas that go beyond sensory experience, but it fails to communicate the meaning of such ideas as it is determined by the various causal and explanatory relationships between different introspective, emotional and affective properties that lie at the background of such ideas.

5

The aesthetic thesis of Kant's cognitivism

I have discussed thus far the epistemic thesis of aesthetic cognitivism within Kant's theory of art as expression of aesthetic ideas. I argued that works of art have an important cognitive value in that they promote the (objectual) understanding of rational ideas and abstract concepts. Yet, it is one thing to claim that works of art have cognitive value and another that such cognitive value is aesthetically relevant. To recall, aesthetic cognitivism is a view claiming not only that works of art serve as a source of knowledge and understanding about the world, but also that such cognitive value determines work's aesthetic value. That is, artwork's cognitive value is one of the reasons that figures in our evaluation of the work as aesthetically successfully. Aesthetic cognitivism, however, does not maintain that all works of art that have cognitive value are thus aesthetically better, that is, it allows the possibility that works of art promote cognition, without this having to do anything with their aesthetic value. Thus, while we may acquire some medical and historical truths from engaging with the television show *The Knick* (2014), the work is not aesthetically better because of its accurate historical portrayal. It is therefore perfectly compatible with aesthetic cognitivism to have aesthetically good works of art that advance false views (an example is Luc Besson's film *Lucy*, which entertains false thematic statement that human beings use only 10 per cent of their mental capacity, but which does not affect the aesthetic worth of the work) and the other way around, aesthetically bad works that display profound truths. Cognitive value of artworks does not necessarily affect their aesthetic value. Rather, as pointed out previously, cognitive value is aesthetically relevant when it is furthered by work's aesthetic properties, such as formal and stylistic properties, narrative details, sensory and representational properties. It is the manner or the way the cognitive insights are conveyed that makes them aesthetically relevant.

According to my account argued thus far, cognitive value of artworks lies in the promotion of objectual understanding, that is, in comprehending the internal relationship between various introspective, emotional and affective properties that to a large extent constitute the meaning of abstract concept and rational ideas. What requires further investigation is specifying the exact nature of the relationship between artwork's promotion of objectual understanding and its aesthetic appreciation. I begin the first part of this chapter by exploring the aesthetic thesis of Kant's cognitivism. While Kant's text appears to endorse two inconsistent views regarding the relationship between artwork's cognitive and aesthetic value, I propose to resolve this inconsistency by distinguishing between two kinds of beauty, perceptual and spirited beauty (beauty of expression of aesthetic ideas), both depending on formalistic criteria of beauty alone. I show that Kant holds an extreme form of aesthetic cognitivism in respect to art, namely, the view that all works of art have aesthetic value (partly) in virtue of their cognitive value. In conclusion, I introduce the idea, fully specified in the last chapter, that feeling of pleasure, which determines the aesthetic value of an artwork, has a cognitive function in that it serves not only as a means by which we come to grasp the objectual understanding, but also as a felt sense that makes such understanding possible.

Kant's aesthetic cognitivism

Kant first introduces the idea that works of art must exhibit a significant cognitive value in terms of expressing aesthetic ideas in §49, where he states the following:

> One says of certain products, of which it is expected that they ought, at least in part, to reveal themselves as beautiful art, that they are without spirit, even though one finds nothing in them to criticize as far as taste is concerned. A poem can be quite pretty and elegant, but without spirit. A story is accurate and well organized, but without spirit. A solemn oration is thorough and at the same time flowery, but without spirit. Many a conversation is not without entertainment, but is still without spirit; even of a woman one may well say that she is pretty, talkative and charming, but without spirit.
>
> (KU 5:313; 191–2)

If an object is produced with an intention to be a work of art, then it must exhibit not only aesthetic value (beauty), but spirit as well, the latter being 'the faculty for the presentation of aesthetic ideas' (KU 5:314; 192). As pointed out previously, aesthetic ideas have a function of connecting the subjective aspects of our abstract and rational concepts with imaginative representations, thereby promoting (objectual) understanding of such abstract phenomena.

Thus, Kant's notion of spirit is essentially construed as a carrier of artwork's cognitive value.

However, while this passage clearly supports the epistemic thesis of aesthetic cognitivism, that is, the idea that works of art must express cognitively valuable aesthetic ideas, it leaves unclear the exact nature of the relationship between artwork's cognitive and aesthetic value. Does spirit, as the faculty of producing cognitively valuable aesthetic ideas, determine artwork's aesthetic value or not? As it stands, the passage supports both readings. For example, one can read Kant's statement that spiritless works of art (i.e. works that produce no cognitive effects), which 'ought, at least in part, to reveal themselves as beautiful', are not actually beautiful, implying thereby that spirit is necessary for the artwork's beauty or aesthetic value. This interpretation is furthermore supported by the fact that Kant uses substantive aesthetic terms, such as 'pretty', 'elegant', 'well organized' and 'charming', to describe spiritless works of art, rather than verdictive or evaluative aesthetic term such as beauty (or aesthetic merit).[1] That is, spiritless artworks can exhibit some aesthetic properties, but they do not possess aesthetic merit (or beauty). It appears accordingly that expression of an aesthetic idea necessarily determines artwork's aesthetic value, the view, which moreover coheres with Kant's claim that all beauty (artistic and natural beauty), 'can in general be called the expression of aesthetic ideas' (KU 5:320; 197). Indeed, Kant often associates beauty, both artistic and natural, with the expression of an aesthetic idea. He writes that trees express the ideas of majesty and magnificence, fields the idea of joyfulness (KU 5:354; 228) and the bird's song the idea of 'contentment with its existence' (KU 5:302; 181). Statements such as these have led many Kant's scholars to interpret the notion of spirit as a necessary element of beauty (Chignell 2007, Rogerson 2009, Rueger 2008). For example, Kenneth Rogerson writes that aesthetic appreciation always 'involves our interpreting a manifold as organized in a way to best express an aesthetic idea' (2009: 21). Something similar is suggested by Alexander Rueger, when he writes that 'not every object appears beautiful […] Rather it is only […] when the given form serves not only as a presentation of the (determinate) concept of the object but also happens to present symbolically, as an aesthetic idea, a different, indeterminate concept' (2008: 311). Both natural and artistic objects appear to possess aesthetic value in virtue of their ability to express aesthetic ideas. Object's cognitive value determines its aesthetic value.

Yet, on the other hand, the passage also supports the interpretation of artistic beauty as being independent of spirit. For instance, Kant writes that certain works of art can be spiritless (i.e. have no cognitive value) 'even though one finds nothing in them to criticize as far as taste is concerned'. Given Kant's definition of taste as the 'faculty for the judging of the beautiful' (KU 5:203; 89), furthermore, his claim that the beautiful 'requires only taste' (KU 5:311; 189), the passage supports the conclusion that spirit is not required for artwork's beauty. That is, works of art can have aesthetic value, without having cognitive value inherent in the expression of an aesthetic idea. Such reading is also consistent with Kant's restatement of the above passage in his *Anthropology*, where he writes: 'One says that a speech, a text, a woman in society, etc., are beautiful but without spirit' (Anthr 7:225; 120). Here Kant clearly uses a verdictive or evaluative aesthetic term 'beauty' to describe objects that have taste, yet lack spirit, suggesting thereby that beauty as the aesthetic value par excellence can exist independently of the spirit embodied in the expression of aesthetic ideas. Indeed, Kant's notion of free beauty, that is, beauty of an object in virtue of its perceptual form alone without considering the concept of a purpose, supports such interpretation. While Kant's typical examples of free beauty are objects of nature, he also includes certain types of artworks and artefacts into this category as well, such as pure instrumental music or wallpapers, which presumably 'signify nothing by themselves: they do not represent anything, no object under a determinate concept, and are free beauties' (KU 5:229; 114). Presumably, we can find some works of art beautiful simply in virtue of their arrangement of perceptual features, without having in mind any ideas or meanings that they might be associated with.

We can see accordingly that Kant's text appears to endorse two seemingly incompatible views regarding the relationship between artwork's cognitive and aesthetic value. On the one hand, there is strong textual evidence in support of an extreme form of aesthetic cognitivism within Kant's aesthetics, namely, the view that all works of art must possess cognitive value (that is, express aesthetic ideas) in order to be aesthetically meritorious. Yet, on the other hand, he appears to be committed to the view that works of art can have aesthetic value independently of possessing cognitive value (that is, to express aesthetic ideas). Presumably, some works of art can be aesthetically appreciated in virtue of their perceptual form alone, without being required to give rise to ideas that go beyond sensory experiences (rational ideas and abstract concepts). Among Kant's scholars, this inconsistency is often presented in a form of a conflict between two different criteria of beauty that Kant seemingly holds, namely, formalistic criterion according to which it is the perceptual form of the object alone that carries an aesthetic quality and an expressive criterion, which takes expressive content, rather than perceptual form as being responsible for object's aesthetic value (Gotshalk 1967, Guyer 1977, Yanal 1994).

My aim in what follows is to resolve this apparent inconsistency within Kant's texts by proposing a distinction between perceptual beauty and spirited beauty (that is, beauty of an aesthetic idea), while arguing that both depend on formalistic criteria of beauty alone. That is to say, I contend that aesthetic form of the object (natural or artistic) can be thought to exists at two levels – on the perceptual level referring to the combination of object's perceptual features and on the level of aesthetic ideas whereby it is constituted by the combination of aesthetic attributes or associational thoughts. Thus, spirited beauty supervenes on formal, rather than representational or semantic properties as well. I find support for this view in Kant's claim that aesthetic form alone functions as the 'vehicle of communication' of aesthetic ideas (KU 5:313; 191), which is 'purposive for observation and judging, where the pleasure is at the same time culture and disposes the spirit to ideas' (KU 5:326; 203). Artistic expression of an aesthetic idea requires not only artist's ability to find aesthetic attributes for a given concept (rational idea or abstract concept), but also the ability to judge objects by means of taste, that is, to find the aesthetic form that exhibits free harmony between cognitive faculties, since it is only in virtue of this harmony that aesthetic ideas can be communicated to others. Taste or experience of beauty is required for the expression of an aesthetic idea that makes sense to others and in virtue of which sensible presentation of concepts can take place. This is the thesis that I will argue for in what follows. In particular, I will attempt to show that one and the same object can have either or both perceptual and spirited beauty (or ugliness), yet, that if an object is an artwork, then it can only have spirited beauty as its appropriate aesthetic value. Thus, according to my interpretation, Kant holds extreme form of aesthetic cognitivism view in respect to art, according to which all works of art must have cognitive value in virtue of expressing aesthetic ideas and this cognitive value necessarily determines work's aesthetic value. All works of art are aesthetically good in virtue of their cognitive value (whereas this cognitive value can be exercised in different degrees, as pointed out in the previous chapter). I explain aesthetically relevant cognitive value by referring to Kant's notion of taste, that is, experience of free harmony (or beauty). I argue that taste (or the aesthetic feeling of pleasure) functions not only as a means by which we come to recognize the attainment of free harmony in the expression of an aesthetic ideas, but also as a rule that provides unified meaning and sense to the manifold of aesthetic attributes (or associational thoughts).

Kant's concept of beauty as purposiveness of form

According to Kant's aesthetic theory, an object has an aesthetic value (beauty) due to its form alone independently of the content of the object. Kant explains the aesthetic value of the mere form of an object by means of his notion of the free harmony between the cognitive faculties of imagination and understanding, the same faculties that are also responsible for our ordinary cognition of objects. To recall, it is Kant's view that our ordinary perception is necessitated by the mental activities of imagination, whose function is to synthesize the manifold of intuition (that is, the form) in order to bring it into an image, and of the understanding, which unifies this manifold under the concept of the object. In synthesizing the manifold of intuition, imagination is not free; rather, it is constrained or subordinated by determinate concepts of understanding, which serve as rules for the imaginative synthesis. Imagination must synthesize the sensible manifold according to the specification of the concept. For instance, in order to recognize a particular object, say a dog, the imagination must follow the dog-rule, that is, it must combine specific features such as a tail, four legs and a head as the dog-rule prescribes. Without this cooperation or harmony between the imagination and understanding there would be no perceptual experience of an object. Kant writes that our perception of the beautiful is based on the same harmonious relationship of cognitive powers, the difference being that in ordinary cognition this harmony is constrained by determinate concepts of the understanding, whereas in aesthetic appreciation, no such concepts restrict imagination and thus their harmony is in free play. Presumably, the imagination has the ability to synthesize the sensible manifold without being governed by any determinate concepts of the understanding, while, however, being in accordance with the general need of the understanding to bring order and unity to the sensible manifold.[2] The experience of such a free harmony between imagination and understanding is distinctive in that it is felt through the feeling of pleasure and in making an aesthetic judgement of the beautiful.

Accordingly, Kant's concept of free harmony appears to involve two constitutive elements, the free play of imagination and the lawfulness of the understanding. The notion of the free play of imagination refers to the ability of imagination to conjure up and combine together the sensible manifold without being governed by any conceptual rules. That is to say, the imagination does not need to synthesize the sensible manifold in light of a particular concept (as it is the case in ordinary empirical perception where the imagination is governed by a determinate rule as to how the combination ought to proceed), but is free to synthesize or combine together sensible manifold in different ways. Kant claims that the subject of aesthetic judgement is the mere form of the object, without the consideration of what the object represents. In other words, the subject of aesthetic judgement is the mere combination of sensible manifold (apprehension), which is not restricted to a particular rule and is therefore free as to how it ought to synthesize the manifold. For example, he writes that pleasure in aesthetic judgement 'is connected with the mere apprehension (apprehensio) of the form of an object of intuition without a relation of this to a concept for a determinate cognition' (KU 5:189; 76). Kant seems to have a view that what we perceive in aesthetic judgements is the combination of sensible manifold (i.e. form) without this form being conceptually determined, that is, without the empirical content of a specific concept. As he writes, aesthetic judgement 'relates the representation by which an object is given solely to the subject, and does not bring to our attention any property of the object, but only the purposive form in the determination of the powers of representation that are occupied with it' (KU 5:228; 113). It is this form alone that carries an aesthetic quality as it is mere apprehension of this form alone that can bring our cognitive faculties in a free harmony that is inherently pleasing.

The subject of aesthetic perception is freely imaginative manifold of intuition, that is, the mere form of the object. Yet, it is not every freely imaginative manifold of intuition that can occasion in us the feeling of aesthetic pleasure. The aesthetic value of the freely imaginative manifold is measured in terms of the degree of the harmonious relation it produces with the faculty of understanding. Even though imagination synthesizes the sensible manifold without being governed by the conceptual rules of the understanding (i.e. imagination is in free play), it must nevertheless stand in agreement with the lawfulness of the understanding. In other words, the combination of the sensible manifold must exhibit a rule-like order, even though this order is not brought up by any particular determinate rule. Kant expresses this form of free harmony between imagination and understanding by referring to his notion of lawfulness without a law: 'Thus only a lawfulness without law and a subjective correspondence of the imagination to the understanding without an objective one – where the representation is related to a determinate concept of an object – are consistent with the free lawfulness of the understanding (which is also called purposiveness without an end)' (KU 5:241; 125). That is, an object is beautiful if the combination of its elements is in harmony with the understanding (it is lawful), but without this harmony being determined by any particular concepts of the understanding (it is without a law). We have the experience of lawfulness without a law when we feel that an object's formal structure is just the right one, in which all elements cohere with each other, without however being governed by any determinate rule that would serve as a basis for the justification of the appropriateness of the specific formal structure. It is the feeling of pleasure alone that expresses the appropriateness of a certain combination of sensible manifold. We simply feel or enjoy the heightened agreement of the faculties of imagination and understanding without this agreement being brought up by the assistance of concepts.

Perceptual and spirited beauty

We know so far that it is one of the central ideas of Kant's aesthetics that mere form of the object has the property of being aesthetically valuable. Kant initially describes the mere form of the object as referring to the synthesized play of sensations (that is, material qualities of an object) in time and space, without conceptual determination. For example, he writes: 'All form of the objects of the senses (of the outer as well as, mediately, the inner) is either shape or play: in the latter case, either play of shapes (in space, mime, and dance), or mere play of sensations (in time)' (KU 5:225; 110). While the mere play of sensations in time refers to the temporal composition of tones (art of music), the play of shapes, on the other hand, denotes the spatial arrangement of lines and figures (i.e. drawing): 'In painting and sculpture, indeed in all the pictorial arts, in architecture and horticulture insofar as they are fine arts, the drawing is what is essential, in which what constitutes the ground of all arrangements for taste is not what gratifies in sensation but merely what pleases through its form' (KU 5:225; 110).[3] On this view, Kant's notion of the mere form of the object refers to the spatial or temporal configurations of perceptual features such as tones, shapes and lines without any conceptual content, that is, what the object is supposed to be or represent (Uehlings 1971, Neville 1974, Berger 2009).

The form of the object can be judged either independently from the concept of a purpose as it is the case in freely beautiful objects or it can be restrained by the concept of the purpose in adherently beautiful objects. Typical examples of free beauty are objects of nature which 'are not attached to a determinate object in accordance with concepts regarding its end' (KU 5:229; 114). For example, the concept of a flower does not determine its purpose for we do not know what a flower ought to be and thus there is no concept of a purpose that would restrain the activity of imagination. Objects of adherent beauty, on the other hand, are works of art and artefacts, which are made with an aim to perform a function of some sort. For such objects, the concept of the object determines their purpose (what they ought to be). Insofar as the concept determines the purpose of the object, it determines the rules for the combination of the manifold (the form of the object). In other words, the concept of the object restrains the free play of imagination. For example, a church is building made with a purpose to worship God. In order to judge the beauty of a church, we must first take into account what the church is and this means to take into account its purpose. In order for the object to be a church, it must fulfil its purpose in the first place. Accordingly, the form of the church is determined by the purpose it is supposed to fulfil, that is, its form must be in accordance with its purpose.

Nonetheless, even though the concept of the purpose restricts the free play of imagination to some degree, it does not restrict the harmony (or disharmony) between the free imagination and the understanding. That is to say, the concept of the purpose does not preclude the free play of imagination completely and therefore it does not preclude free harmony (or disharmony). There are numerous different forms that satisfy the purpose of the church; yet, not all of them are beautiful. The beauty (or ugliness) of a church is not determined by the satisfaction of the purpose, even though it depends on it. The satisfaction of the purpose of the church is a necessary, but not a sufficient, condition of its beauty (or ugliness). Within the constraint of the purpose, the imagination has an ability to play freely with the given sensible manifold and therefore allows for the possibility of free harmony (or disharmony). The consideration of the purpose of the object restricts the range of the appropriate forms, that is, it restricts the freedom of the play of imagination, but it does not preclude it completely.[4] And as long as in the apprehension of a given object the imagination can be free to some extent, genuine judgement of taste, based on free harmony (or disharmony), can be given.

However, if it is the form of the object, either unrestrained or restrained by the concept of the purpose that carries the aesthetic value of an object, then how can Kant ascribe beauty to work's expressive content that presumably takes place in artistic expression of aesthetic ideas? That is to say, how to account for an apparent change in Kant's text from the view that beauty is a reflection of object's form (i.e. spatial or temporal configurations of perceptual features without any representational content) to his claim that beauty consists in the presentation of concepts that go beyond sensory experience, that is, in the expression of aesthetic ideas?

According to D. W. Gotshalk, the answer is simple – Kant holds two different, incompatible theories of beauty. On the one hand, he holds a formalist theory of natural beauty, which focuses exclusively on the perceptual form as the criteria of beauty and an expressionist theory of artistic beauty, according to which it is the expressive content, rather than the form that produces aesthetic pleasure. As the author writes, in Kant's theory of art 'expression rather than form becomes the universal principle of Beauty' (1967: 253).

Gotshalk's division between the two theories of beauty has been convincingly criticized by Paul Guyer. He points out that Kant often refers to beauty of artworks in strict formalistic terms (for instance, in §14 Kant discusses form as the sole basis of aesthetic pleasure in paintings and sculptures), whereas he describes beauty of natural objects in expressionistic terms (trees expressing the idea of majesty or fields the idea of joyfulness). Based on this, Guyer concludes that Kant offers one, albeit complex theory of beauty: 'Kant's formalism and expressionism may be seen as aspects of a complex but non-contradictory theory of the pleasures which we take in beautiful objects' (1977: 48). He argues that the apparent conflict between Kant's initial commitment to the view that mere perceptual form serves as a sole source of aesthetic pleasure and his latter view that semantic and conceptual content in the expression of aesthetic ideas can also dispose the mind to free harmony is a result of 'Kant's mistake in interpreting his doctrine of formalism' (1977: 57), but which does not significantly undermine the coherency of Kant's aesthetic theory. Presumably, Kant's concept of free harmony does not by itself exclude any role of representative and expressive properties as being a part of the manifold that can engage our aesthetic attention and contribute to aesthetic experience. As he writes:

> Nothing in this doctrine need be seen as excluding concepts, or representations of concepts, symbols, and the like, from being part of the manifold of Imagination which the mind ranges over in its free play. Nothing in what I have argued to follow so far from the theory of cognitive harmony need exclude the meaning or significance, the suggestiveness or symbolic aptness, of a given representation, work of art or of nature, from being among that which disposes the mind to the state which grounds an aesthetic judgment. Thus, there does not seem to be any prima fade reason why Kant's epistemological analysis of the aesthetic response should be incompatible with the claim that the beautiful gives rise to aesthetic ideas.
>
> (1977: 55)

Guyer is persuasive in demonstrating that Kant holds one, rather than two different accounts of beauty. He resolves the tension between formalism and expressionism by expanding Kant's concept of beauty (or free harmony) to include not merely formal, but representational and expressive properties as well. However, by making such a move he substantially undermines the role that formalism plays in Kant's theory of aesthetics. On my view, such a restrictive solution is not necessary. Instead of expanding Kant's conception of beauty or free harmony, it suffices to expand his conception of formalism to include not merely perceptual, but representational and semantic properties as well.[5] That is to say, Kant's conception of aesthetic form, which initially refers to the organization of object's perceptual features, need not exclude the organization of semantic and intellectual elements in an aesthetic idea. Both perceptual beauty and spirited beauty (i.e. beauty of an expression of aesthetic idea) depend on formalistic criteria of beauty alone, that is, in both cases, it is the formal arrangement of either perceptual or semantic properties that brings our cognitive faculties into a free harmonious play. As pointed out previously, an aesthetic idea has a formal structure as it is composed of a plurality of mental representations or aesthetic attributes. These aesthetic attributes constitute the content (material) of the aesthetic idea; yet, it is their form, that is, the organization of aesthetic attributes, that generates the expression of an aesthetic idea and produces the attunement of cognitive faculties. Something similar is suggested by Diarmuid Costello when he writes that aesthetic form can be constituted by 'the unified organization of aesthetic attributes required to present an idea [...] The way in which a work organizes and unifies its aesthetic attributes so as to convey an idea just is its "form" in this expanded sense' (2013: 18). A related view is purported by Andrew Chignell as well: 'It is not the content of these thoughts' but rather 'the formal manner in which these thoughts are strung together by the mind into a "coherent whole" that [...] is essential to the experience of an aesthetic idea' (2007: 424). Construed this way, Kant's claim that beauty is an expression of aesthetic ideas remains compatible with his doctrine of formalism; beauty is the organization of aesthetic attributes (semantic material) that gives rise to an aesthetic idea.

In this respect, my interpretation remains faithful to the central thesis of Kant's aesthetics, namely that aesthetic pleasure is a disinterested pleasure due to the formal qualities of an object alone and which is applied both to natural and artistic objects. As Kant writes in his discussion of artistic beauty, 'So much for the beautiful representation of an object, which is really only the form of the presentation of a concept by means of which the latter is universally communicated' (KU 5:312, p. 191). Aesthetic value of an artwork resides not in the particular idea or thought communicated by the object, but rather in the particular way these ideas and thoughts are arranged together to form a coherent unity, since it is this unity that produces aesthetic pleasure. Accordingly, the value of both perceptual and spirited beauty is measured in terms of the feeling of pleasure induced by the form of an object, whereby this form is constituted either by the manifold of perceptual features or by the manifold of aesthetic attributes through which aesthetic ideas are expressed. Expressive power of spirited beauty resides in its formal excellence as it is the form alone that is responsible for communicating an aesthetic idea. Thus, both perceptual beauty and spirited beauty conform to the general formalist theory of beauty according to which it is the form, rather than the content, that carries aesthetic value.

Distinction between perceptual and spirited beauty (or beauty of aesthetic ideas) can account for many of the challenges facing Kant's aesthetic theory of art (or aesthetic theory of art in general).[6] One of such challenges can be formulated as following: if value of artworks lies exclusively in their possession of pleasure-inducing aesthetic properties (beauty), whereas all aesthetic properties supervene on perceptual-formal properties (that is, physical and sensory properties such as arrangement of lines, shapes, colours, sound), then non-perceptual works of art, such as literary and conceptual art, which do not depend on any perceptual properties, but rather in the ideas, concepts and meanings they evoke, cannot have artistic value. Such works of art presumably refute an aesthetic theory of art.[7] On my interpretation of Kant's formalism, non-perceptual artworks can be easily accommodated within aesthetic theory of art. While such artworks do not possess perceptual beauty for they lack perceptual properties (literary art) or they lack any significant perceptual properties (conceptual art has a physical form, but which is irrelevant for its artistic value), they can nevertheless possess spirited beauty, that is, they can be aesthetically valuable due to their particular formal structure of aesthetic attributes (associational thoughts, feelings, emotions) that yields an aesthetic idea. As pointed out previously, an aesthetic idea is significantly different from the empirical intuition (perceptual image) in that it is an internal, rather than external, representation of imagination – a mental image of some sort. Furthermore, while empirical intuition is constituted by the synthesis of perceptual features, which can bring our faculties in a free harmonious (or disharmonious) play, an aesthetic idea is constituted by the synthesis of associational thoughts (or aesthetic attributes) and their combination can as well exhibit free harmony (or free disharmony). This suggests that one and the same object can have either or both perceptual beauty (or ugliness) and spirited beauty (or ugliness).[8] An object can be aesthetically valuable not merely due to its visual form alone, but also because of the aesthetic idea it communicates to the audience.[9] Hence, non-perceptual art can be aesthetically appreciated in virtue of its aesthetic-formal properties.

Another challenge brought up against the aesthetic theory of art is that presumably it cannot explain our appreciate powers underlying difficult works of art that evoke painful and negative aesthetic experiences, yet, which we evaluate as aesthetically valuable and captivating. If the value of artworks is measured in terms of the aesthetic feeling of pleasure in the beautiful, then ugliness, which occasions in us the feeling of displeasure, must be associated with aesthetic disvalue. There is no possibility to accommodate within this aesthetic picture works of art, prevalent in contemporary artistic production, that evoke (and aim to evoke) negative aesthetic feelings of ugliness and repulsion, and the positive appreciation of them. In other words, Kant's aesthetic theory of art cannot account for the so-called 'paradox of ugliness', namely, how we can like, attend to and value something that we prima facie do not like, find positively displeasing or even repellent? Presumably, ugliness itself merits aesthetic significance and is often characterized by contemporary writers as fascinating, astonishing and captivating (Lorand 1994, Kieran 1997, Brady 2010). Indeed, if we take a closer look at the Willem de Kooning's painting *Woman I* (1950–2), we can notice that the painting captivates our attention precisely for the same reason it repels us, namely, due to the grotesque disfiguration of the image of the women's body. We can perceive the depicted woman's invasive breasts, bulging eyes, teeth spreading into a grinning smile, while the rest of the body – her arms and torso – is disintegrated, dismembered and dissolved into the spontaneous and dynamic brush strokes, with frantic lines and garish colours. De Kooning's painting nicely illustrates the paradoxical character of ugliness, namely, that we can still find aesthetic value in looking at an object that we visually dislike. The painting is captivating and holds our attention precisely because of those features that cause discomfort in the first place. How can such concurrence of aesthetic displeasure and positive aesthetic appreciation be explained?

I argue that this phenomenon can be explained by means of the proposed distinction between perceptual beauty (or ugliness) and spirited beauty (or ugliness). For example, while the perceptual form of De Kooning's painting is itself chaotic and displeasing (perceptual ugliness), it can still be experienced as thoughtful and aesthetically significant due to its expression of an aesthetic idea (spirited beauty). It is precisely the discomforting perceptual features of the women's body, composed together in a disintegrated way, that serve as aesthetic attributes giving rise to the idea, such as the idea of a critique of a social, aesthetic and moral idealization of femininity. This idea is suggested by the violence of the brushstrokes, the chaotic and aggressive combination of colours, the idea of sexual dominance expressed through the accentuation of the women's breasts, and the maliciousness, hostility and pretence conveyed by her grinning smile. The expression of this idea is stimulating, thought-provoking, and for this reason aesthetically significant and valuable, even though the perceptual form is itself displeasing and ugly. Through the unique representation of a woman, the artist managed to express an idea which cannot be represented otherwise, that is, he succeeded to express an aesthetic idea, and this in itself is a valuable experience, even though the resulting work is visually ugly. Even non-representational art, while perceptually displeasing, can be aesthetically valuable due to the expression of aesthetic ideas. Consider, for example, conceptual musical work *Imaginary Landscape No.2* (1942) by John Cage. This piece is composed of various sounds produced by unconventional instruments, such as tin cans, buzzers, water gongs and conch shells. The combination of these sounds produces a raucously noisy and chaotic work; it lacks melody, harmony and organization, and it is therefore difficult to listen to. Yet, it is precisely through its incongruous combinations of sounds that the artist is able to explore the ideas of chance, unpredictability and absurdity inherent in one's desires to bring order and meaning into the unpredictable world. There is an appealing side to perceptual ugliness, because it allows for the expression of ideas that cannot be represented otherwise. In particular, given that the constitutive element of ugliness is disorder, it is particularly suggestive for the expression of ideas that celebrate such disorder, such as ideas of alienation, estrangement, dehumanization, destruction, degeneration, disconcertion, absurdity, horror, anxiety and fear.

Art as spirited beauty

As argued thus far, an object can have aesthetic value in virtue of either or both perceptual and spirited beauty (or beauty of an expression of aesthetic ideas). What I want to demonstrate next is that if an object is made with an intention to reveal itself as a beautiful artwork, then it can do so only by means of exhibiting spirited beauty. According to Kant, artwork's aesthetic value essentially resides in its ability to express cognitively valuable aesthetic ideas. As he writes, 'a product of beautiful art requires not merely taste, which can be grounded on imitation, but also originality of thought, which, as self-inspired, is called spirit. […] the painter of ideas alone is the master of beautiful art' (Anthro 7:248; 145). Kant distinguishes here between artworks that exhibit mere taste (or mere perceptual beauty) and artworks with taste and spirit (spirited beauty). An example of the former is an artwork that has aesthetic value in virtue of their perceptual beauty alone and which can be displayed by imitating visually beautiful nature. For instance, Thomas Moran's landscape painting *Shepherdess Watching Her Flock* (1857) merely imitates the perceptual beauty of nature (the beauty of a painting supervenes on the beauty of nature), without expressing any ideas. On the other hand, spirited art paints aesthetic ideas in the sense of sensibly representing concepts that go beyond sensory experience and thus imbues the work with a significant cognitive value. Spirited art alone, as Kant writes in the above passage, deserves to be called beautiful art.

Accordingly, the aesthetic value of an artwork is measured in terms of the degree to which it disposes the mind to the 'fullness of thought to which no linguistic expression is fully adequate, and thus elevates itself aesthetically to the level of ideas' (KU 5:326; 204). As argued previously, literary art, which exhibits spirited beauty alone, concurs in this respect as it is not tied down by empirical intuition and can thus offer a 'mere play with ideas' providing thereby 'nourishment to the understanding in play, and giving life to its concepts through the imagination' (KU 5:321; 199). While pictorial art depends on the appeal of the senses and thus exhibit perceptual beauty (or ugliness), its aesthetic value is primarily measured in terms of making 'shapes in space into expressions of ideas' and 'the mere expression of aesthetic ideas is the chief aim' (KU 5:322; 199). Even non-representational art, which consists in mere perceptual form, must obey the principle of spirit and connect the play of sensations (tones, colours) with semantic content in order to be valued as beautiful art.

Kant's insistence on identifying artistic beauty with spirited beauty (or beauty of an expression of aesthetic ideas) can be explained by referring to his notion of beauty as purposiveness without purpose (or free harmony). That is, harmony in the given object must be attained freely, without being determined by the concept of the purpose with which the object is produced. Kant writes that to judge the object based on the concept of a purpose is to make a judgement of perfection, rather that one of taste (KU 5:241; 125). Prima facie, the notion of free harmony as applied to art seems problematic, considering Kant's claim that to judge the beauty of an artwork 'one must be aware that it is art, and not nature' (KU 5:306; 185). That is to say, in judging the beauty of an artwork we must take into account the purpose of the object (what it ought to be) and hence 'the perfection of the thing will also have to be taken into account, which is not even a question in the judging of a natural beauty (as such)' (KU 5:311; 190). Artistic beauty presupposes the concept of a purpose and it is therefore beauty of the adherent kind: 'For something in it must be thought of as an end, otherwise one cannot ascribe its product to any art at all; it would be a mere product of chance' (KU 5:310; 188). That is, the organizational structure of an artwork is not accidental, but is made in accordance with a certain purpose in the artist's mind. This means that there are certain rules that guide the artist in creating their work: 'For every art presupposes rules which first lay the foundation by means of which a product that is to be called artistic is first represented as possible' (KU 5:307; 186). Yet, even though production of art presupposes certain rules, Kant demands that 'the purposiveness in its form must still seem to be as free from all constraint by arbitrary rules as if it were a mere product of nature' (5:306; 185). In other words, in order to judge the beauty of an artwork, one must be aware that the object is an artwork and thus created for some purpose, and therefore in accordance with some rules. Yet, these rules cannot be of a determinate kind: 'It cannot be couched in a formula to serve as a precept, for then the judgment about the beautiful would be determinable in accordance with concepts' (KU 5:309; 188). The purposiveness in an artwork must be free of rules, as if a product of a spontaneous and accidental activity.

The notion of spirited beauty satisfies both seemingly incompatible criteria that Kant places on beautiful art. Namely, spirited beauty depends on the concept of a purpose with which the object is created, that is, to sensibly represent concepts that go beyond sensory experience (such as rational ideas or abstract concepts) and therefore it is made in accordance with certain rules. However, it is distinctive for such concepts that no determinate rules can be given for them. For example, one does not know what the ideas of loneliness, truth, justice, heaven, etc., ought to look like. What is distinctive about such ideas is that they have no appropriate empirical intuition. In comparison to determinate concepts (such as a concept of a flower, a dog, a table, etc.) that have a direct empirical counterpart, ideas such as loneliness, freedom, justice, happiness have no particular shape, size or colour, and one cannot see, touch or hear them. Consequently, one cannot provide determinate rules in accordance with which to produce the sensible manifold for them. This means that the manifold of aesthetic attributes, which constitutes an aesthetic idea, is combined together freely, without being governed by any determinate rule as to how the combination ought to proceed: 'The aesthetic idea is a representation of the imagination, associated with a given concept, which is combined with such a manifold of partial representations in the free use of the imagination' (KU 5:316; 194). Thus, even though artistic expression of an aesthetic idea depends on the concept of a purpose (this concept being abstract concept or rational idea), the play of imagination and understanding can nevertheless be as free as in the case of natural objects. This is, in a nutshell, the idea that Kant has in mind when he says that 'art can only be called beautiful if we are aware that it is art and yet it looks to us like nature' (KU 5:306; 185). That is, artworks are unlike natural objects in that they depend on the concept of a purpose; yet, they are like natural objects in that no determinate rules for the combination of the manifold can be given. Kant claims it is the nature in genius that gives the rules to art: 'Nature in the subject (and by means of the disposition of its faculties) must give the rule to art, i.e., beautiful art is possible only as a product of genius' (KU 5:307; 186). Genius is 'the talent (natural gift) that gives the rule to art', this being 'a talent for producing that for which no determinate rule can be given' (KU 5:307; 186). The production of an artwork is not governed by any known rules; rather, it is the artist that creates the rule for the combination of artistic material. In creating the new rule, the artist is governed by his nature alone, and this nature is the ability to exercise the free play of his cognitive powers: 'Genius is the exemplary originality of the natural endowment of a subject for the free use of his cognitive faculties' (KU 5:318; 195). Thus, while artistic beauty depends on the concept of a purpose, its purposiveness is nevertheless the result of the same freedom in the play of cognitive powers that one can recognize in judging the beauty of nature.

According to Kant, beauty of an artwork depends on its ability to express aesthetic ideas (i.e. have spirited beauty). While we can aesthetically appreciate work of art based on its perceptual form alone (perceptual beauty), such an appreciation is not an appropriate one. In order to give an appropriate aesthetic judgement regarding a certain object, we must take into account what the object really is. If the object is an artwork and thus intentionally produced for a certain purpose, i.e. the purpose of representing ideas that go beyond sensible experience, then we must take this purpose into account when judging its aesthetic value. Judging the beauty of the artwork therefore presupposes the knowledge of the work's purpose which determines its existence as an artwork. Thus, judging the artwork based on its perceptual form alone, without considering the purpose that governs the production of the work, is judging it inappropriately. The case is different with judging the beauty of natural objects as they do not presuppose any concept of purpose. We do not know, for example, what the flower ought to be and thus there is no concept of a purpose to take into account in judging the beauty of the flower. Even though, as Kant writes, the botanist knows the natural purpose of the flower, namely, that it has a biological function as the plant's organs of reproduction, he 'pays no attention to this natural end if he judges the flower by means of taste' (KU 5:229; 114). Presumably, while we might perceive natural objects in light of their natural purposes, these purposes have been nevertheless deployed by us to serve our need to systematize nature for the sake of understanding it, whereas nature would exist regardless of our intentions to attribute purposes to it. Thus, natural objects can have only perceptual beauty as its appropriate aesthetic value. While Kant often associates natural beauty with the expression of aesthetic ideas, such beauty cannot be strictly speaking attributed to natural objects as they lack the intentionality that is built in the notion of spirited beauty. Spirit, according to Kant, is the talent of the genius for the presentation of aesthetic ideas, which 'purposively sets the mental powers into motion' (KU 5:313; 192). Natural objects can be seen as expressing aesthetic ideas only when we think of them as if intentionally produced (by an unknown source) to suit us and please us: 'The flowers, the blossoms, indeed the shapes of whole plants; the delicacy of animal formations of all sorts of species, which is unnecessary for their own use but as if selected for our own taste [and] which seem to have been aimed entirely at outer contemplation' (KU 5:347; 222). By itself, however, nature cannot express aesthetic ideas. As Paul Guyer writes, 'the beauties of nature have only the semblance of expressiveness, and are not actually to be regarded as expressing anything' (1977: 69).

Relationship between cognitive and aesthetic value

According to Kant, value of artworks lies in their ability to express aesthetic ideas. An aesthetic idea, as argued previously, is similar to empirical intuition in that they are both product of the synthesis of imagination. Yet, whereas an empirical intuition is a product of the imaginative synthesis of various sense impressions, an aesthetic idea, as an inner (mental) picture, is a product of the imaginative synthesis of various thoughts and associations (i.e. aesthetic attributes). Furthermore, while in the case of an empirical intuition the synthesis of imagination is governed by a determinate concept (in order to recognize a particular object, say as a flower, the imagination must follow the rule as specified by the concept of a flower), in the case of an aesthetic idea, the synthesis of imagination is not determined by any rule and thus it is in a free play: 'The aesthetic idea can be called an inexponible representation of the imagination (in its free play)' (KU 5:343; 218). On Kant's view, it is the experience of this freedom in the play of imagination and understanding that produces the aesthetic feeling of pleasure in the beautiful (when cognitive faculties are in free harmony) or displeasure in the ugliness (when cognitive faculties are in free disharmony).

However, it is not every synthesis or combination of aesthetic attributes that can express an aesthetic idea. Rather, an aesthetic idea is communicated to others when the combination of aesthetic attributes (that is, set of associational thoughts, feelings and emotions) shows its agreement with taste or aesthetic reflective judgement. Taste is a critical faculty of judging 'the suitability of the thing (of its form) to our cognitive faculties' (KU 5:194; 79) by means of the feeling of pleasure. It is by means of taste that we are able to experience the playful interaction between different aesthetic attributes that is not governed by any determinate rules, but which nevertheless exhibits harmony and coherent unity as required for the expression of an aesthetic idea. This is, in a nutshell, the idea Kant has in mind when he says that 'genius really consists in the happy relation [...] of finding ideas for a given concept on the one hand and on the other hitting upon the expression for these, through which the subjective disposition of the mind that is thereby produced, as an accompaniment of a concept, can be communicated to others' (KU 5:317; 194).

An aesthetic idea by itself is merely a product of the free play of imagination offering a rich and novel collection of aesthetic attributes or associational thoughts. But in order for such an aesthetic idea to be intelligible it must be combined with the faculty of understanding: 'for to express what is unnameable in the mental state in the case of a certain representation and to make it universally communicable, whether the expression consist in language, or painting, or in plastic art – that requires a faculty for apprehending the rapidly passing play of the imagination and unifying it into a concept' (KU 5:317; 195). To express an aesthetic idea, both productive imagination and taste are required. Productive imagination is responsible for producing the manifold of associational thoughts for which no determinate concept can be given (i.e. freely imaginative manifold), whereas function of taste is to bring coherence into the given manifold of associational thoughts. As Kant writes, 'Spirit and taste: spirit to provide ideas, taste to limit them to the form that is appropriate to the laws of the productive power of imagination' (Antro 7:246; 144). Taste is responsible for choosing and recognizing the kind of aesthetic form that provides sense and meaning to the collection of aesthetic attributes:

> Taste, like the power of judgment in general, is the discipline (or corrective) of genius, clipping its wings and making it well behaved or polished; but at the same time it gives genius guidance as to where and how far it should extend itself if it is to remain purposive; and by introducing clarity and order into the abundance of thoughts it makes the ideas tenable, capable of an enduring and universal approval, of enjoying a posterity among others and in an ever progressing culture.
>
> (KU 5:319; 197)

The formal structure of aesthetic attributes must be internally coherent and purposive for the presentation of a given rational or abstract concept (similarly as the combination of logical attributes in an empirical intuition must be in a certain combination in order to be recognized as representing a certain empirical concept). The function of taste is to ensure the attunement between aesthetic attributes in order to exhibit internal order and unity. If aesthetic attributes are not combined together in this way, the result is a meaningless and nonsensical representation, or what Kant calls 'original nonsense' (KU 5:308; 186).[10] Mere aesthetic ideas without taste leads to original presentations (novel images, combinations of words and sounds), which carry no intelligible meaning:[11]

> To be rich and original in ideas is not as necessary for the sake of beauty as is the suitability of the imagination in its freedom to the lawfulness of the understanding. For all the richness of the former produces, in its lawless freedom, nothing but nonsense; the power of judgment, however, is the faculty for bringing it in line with the understanding.
>
> (KU 5:319; 197)

We know so far that taste or experience of free harmony between our cognitive powers of imagination and understanding is required for artistic expression of an aesthetic idea since it is this harmony that brings sense and meaning to the combination of aesthetic attributes through which (objectual) understanding of abstract phenomena can be achieved. Furthermore, that experience of such free harmony has always 'the pleasure in the object as a consequence' (KU 5:217; 102), this pleasure being the ground of judging the object as beautiful. This means that artistic expression of an aesthetic idea, which is the carrier of artworks' cognitive value of objectual understanding, always produces the feeling of pleasure and the experience of the beautiful in the audience. In other words, artwork's cognitive value is always accompanied with the feeling of pleasure (aesthetic value).

In fact, the connection between artwork's cognitive and aesthetic value can be pursued even further when considered in light of Kant's claim that the feeling of pleasure is not a mere effect of the free harmony between imagination and understanding, but rather its awareness or recognition. For example, he writes that the 'subjective unity of the relation can make itself known only through sensation' (KU 5:219; 104) or 'this inner relationship [...] cannot be determined except through the feeling (not by concepts)' (KU 5:238–9; 123). We come to recognize free harmony between cognitive powers only by means of the feeling of pleasure. The feeling of pleasure has a recognition role; we feel aesthetically (by means of pleasure) when our cognitive powers are freely harmonious.[12] Now, if free harmony necessitates the expression of an aesthetic idea by providing meaning to the manifold of aesthetic attributes, whereas this harmony is not only the ground of the feeling of pleasure, but rather is itself recognized by means of the feeling of pleasure, then it follows that we come to recognize the expression of an aesthetic idea and its resulting cognitive effect aesthetically, through the feeling of pleasure. In other words, the feeling of pleasure appears to have an active role in the expression of an aesthetic idea in that it registers or detects the attainment of the unified meaning in the combination of aesthetic attributes, that is, associational thoughts, feelings, emotions and other mental aspects that constitute the content of abstract phenomena. We come to apprehend the indeterminate material or multitude of thoughts evoked by the manifold of aesthetic attributes by means of the feeling of pleasure. This idea does appear to be suggested by Kant in §49, where he writes that aesthetic idea is a representation of imagination, 'which is combined with such a manifold of partial representations in the free use of the imagination that no expression designating a determinate concept can be found for it, which therefore allows the addition to a concept of much that is unnameable, the feeling of which animates the cognitive faculties' (KU 5:316, p. 194). Here Kant identifies the 'unnameable', which refers to the multitude of thoughts and meanings brought to mind by aesthetic attributes, with the feeling that animates our cognitive faculties of imagination and understanding, this feeling being nothing else but the feeling of pleasure. Presumably, the mind feels the comprehension of the various introspective, emotional and affective aspects brought to mind by the aesthetic attributes. It moves us aesthetically, by means of the feeling of pleasure, to grasp the (objectual) understanding of abstract concepts and rational ideas.

We can see accordingly that artwork's cognitive value is intrinsically intertwined with its aesthetic value. Without the pleasing-free harmony between cognitive powers of imagination and understanding, no cognitively valuable expression of aesthetic ideas would take place for it is precisely this harmony that provides 'clarity and order into the abundance of thoughts' (KU 5:319; 197). And without the expression of aesthetic ideas and its cognitive effects, there would be no feeling of pleasure. Taste or pleasurable experience of free harmony necessitates the expression of an aesthetic idea by means of providing intelligibility to the manifold of aesthetic attributes (i.e. associational thoughts, feelings, emotions and other mental representations). Presumably, taste functions as a rule of structure that governs the imaginative ordering of the aesthetic material in a specific way, namely, in a way that brings coherent and compelling sense to it. But this suggests that the role of the feeling of pleasure in aesthetic experience of the beautiful appears to be similar to the role that determinate concepts play in cognitive judgement, namely, they provide meaning and sense to the sensible manifold. Kant indeed suggests this idea in §49, where he introduces a distinction between a logical and an aesthetic mode of combining elements and ideas together. He writes:

> There are in general, to be sure, two ways (modus) of putting thoughts together in a presentation, one of which is called a manner (modus aestheticus) and the other of which is called a method (modus logicus), which differ from each other in that the former has no other standard than the feeling of unity in the presentation, while the latter follows determinate principles in this; for beautiful art, therefore, only the first is valid.
>
> (KU 5:318–319; 196)

Logical mode of combining thoughts together depends on a determinate principle, this being a determinate concept that serves as a rule for the unity of sensible manifold in cognitive judgements. Yet, there is also an aesthetic mode of arranging thoughts together where it is the feeling, rather than a determinate concept, that serves as a standard determining the organization of the aesthetic manifold. Here Kant explicitly suggests the possibility that a feeling of pleasure can do the work of actively synthesizing features of an object and thereby guiding our attention in the apprehension of beauty. Accordingly, the role of the feeling of pleasure is twofold; it functions not merely as a means by which we come to recognize a relationship between imagination and understanding as freely harmonious, but also as a felt sense that guides us in ordering the sensible manifold and making free harmony possible. To the elaboration of this idea, I turn next.

6

Kant and aesthetic cognition

In the previous chapter, I introduced the idea that the aesthetic feeling of pleasure serves a role not only of recognizing the free harmony in the combination of aesthetic attributes that yields an aesthetic idea, but also as a felt sense that guides the artist's production (as well as audience's recognition) of such free harmony. If this is the case, then the feeling of pleasure appears to play a role similar to that of a determinate concept; it functions as a rule for bringing meaning and sense to the sensible manifold and for making cognition possible. But this suggests that feeling of pleasure itself has a cognitive function and thus cannot be considered as cognitively irrelevant as Kant initially appears to suggest.

For example, when Kant first introduces the distinction between cognitive judgements and judgements of taste (i.e. judgements of the beautiful) he claims that the former 'relate the representation by means of understanding to the object for cognition', whereas the latter 'relate it by means of the imagination (perhaps combined with the understanding) to the subject and its feeling of pleasure or displeasure' (KU 5:203; 89). A judgement of taste, he writes, is thus 'not a cognitive judgment, hence not a logical one, but is rather aesthetic, by which is understood one whose determining ground cannot be other than subjective' (KU 5:203; 89). Judgements of taste are non-cognitive judgements because they lack conceptual determination. They depend on the harmony between the imagination and understanding that is not determined by the concept of the object in contrasts to cognitive judgements where the harmony is conceptually restrained. Kant explains in the first *Critique* that the function of a concept is to provide meaning and sense to the manifold of intuition (A51/B75). Without concepts applied to the manifold of intuition, all we would be left with is a diverse and confusing array of sense impressions that we would never be able to understand. Hence, to yield cognition, the manifold of intuition must be subsumed under determinate concepts. Since a judgement of taste lacks the essential element that makes cognition possible, that is, the concept, it therefore 'contributes nothing to cognition but only holds the given representation in the subject up to the entire faculty of representation, of which the mind becomes conscious in the feeling of its state' (KU 5:203; 90).

Yet, the idea of judgements of taste having non-cognitive character appears to be less convincing when one turns to Kant's theory of artistic beauty as an expression of an aesthetic idea. To recall, artistic expression of an aesthetic idea is constituted by the following three elements: first, concepts that lack an adequate empirical counterpart (that is, rational ideas and abstract concepts) and which represent the theme of the artwork; second, aesthetic attributes or imaginative associational thoughts that constitute the form of aesthetic ideas and serve as their vehicles; third, the suitability between concepts that lack empirical counterpart and the form, that is, harmony between the freely imaginative synthesis of aesthetic attributes and given concepts. We must experience a sense of freedom in the playful interaction between different associational thoughts that are not conceptually determined, but which nevertheless exhibit order and unity as required by the faculty of understanding. Moreover, Kant writes that free harmony between cognitive powers of imagination and understanding, which we recognize through the feeling of pleasure, necessitates the expression of an aesthetic idea by means of bringing coherent order and sense into the multitude of thoughts. In other words, it is pleasing-free harmony between cognitive powers that provides unified meaning to the manifold of aesthetic attributes or associational thoughts that yields an aesthetic idea. But this suggests that aesthetic feeling of pleasure plays a role similar to that of concepts in determinate cognition, namely, it functions as a rule of structure governing the imaginative organization of aesthetic material in a specific way, namely in a way that brings sense and coherence to it.

This is the idea that I will advance in this chapter. My aim is to argue that in spite of Kant's claim that judgements of taste cannot yield cognition because they do not employ determinate concepts, they must nevertheless be considered as an exercise of some form of cognition, in particular of what Kant calls cognition in general.[1] I explore the distinction Kant makes between determinate cognition and cognition in general by arguing that whereas the former occurs by subsuming the manifold of intuition under the concept of the object, the latter, however, does not necessarily subsume under the concept, but rather under the idea of the empirical world as a systematic whole, that is, under the a priori principle of the purposiveness. A judgement of taste is an exercise of this general sense of cognition by virtue of the principle of purposiveness it necessarily involves and which is manifested through the feeling of pleasure. I demonstrate in what way the feeling of pleasure in judgements of taste substitutes for the role of concepts in determinate cognition. I argue that the feeling of pleasure functions not merely as a means by which we come to recognize a relationship between imagination and understanding as freely harmonious, but also as a felt sense that guides the imaginative ordering of the aesthetic material and making free harmony possible.

The notion of cognition in Kant's aesthetic theory

According to Kant, judgements of taste are non-cognitive judgements due to their non-conceptuality: they are neither grounded on a determinate concept of the object nor do they lead to such a concept (KU 5:209; 95). Judgements of taste are not grounded on the concept of the object in the sense that when we judge aesthetically, say, a flower, the concept of the flower does not impinge upon our reflection. We judge merely its form without considering whether this form satisfies all the features thought in the concept of the flower. Kant writes that a judgement of taste substantially differs from a cognitive judgement in that 'the latter subsumes a representation under concepts of the object, but the former does not subsume under a concept at all' (KU 5:286; 167). The truth or falseness of cognitive judgements such as 'X is a flower' can be proven by rational consideration; the judgement 'this X is a flower' is true if it satisfies the necessary conditions for the application of the concept of a flower. Yet, no such rational consideration is available in the case of a judgement of taste. Whether an object is beautiful cannot be discerned by whether it satisfies the properties of the concept. That is, a given object may be a perfect example of the kind it belongs to, yet be considered ugly.[2] Beauty of an object resides solely in the subject's feeling of pleasure independently of our idea of what the object ought to be. As Kant states:

> If one judges objects merely in accordance with concepts, then all representation of beauty is lost. Thus there can also be no rule in accordance with which someone could be compelled to acknowledge something as beautiful. Whether a garment, a house, a flower is beautiful: no one allows himself to be talked into his judgment about that by means of any grounds or fundamental principles. One wants to submit the object to his own eyes.
>
> (KU 5:215; 101)

Judgements of taste also do not lead to a determinate concept. That is to say, they do not result in a determinate linguistic expression. When we find an object beautiful, we feel there is a tangible account of this, as if beauty were a concept, yet we are unable to put it into words. All that we can say is that the object is beautiful because it elicits in us a certain pleasurable feeling. The experience of pleasure is the sole evidential ground of a judgement of the beautiful.

Kant identifies conceptuality with determinate cognition (bestimmte Erkenntnis) to which he refers to as the 'cognition in the proper sense' (A78/B103). Determinate cognition involves both intuition through which the object is given and concepts through which the object is thought (A50/B74). It is their agreement or harmony that yields determinate cognition: 'The understanding cannot intuit anything, and the senses cannot think anything. Only from their union can cognition arise' (A51/B75). More specifically, determinate cognition arises by means of the harmony between the faculty of imagination, which synthesizes the manifold of intuition in order to bring it into an image, and the faculty of understanding, which unifies the manifold under the concept of the object. Moreover, given that Kant understands concepts as rules for the synthesis of the sensible manifold (A106), this means that concepts are responsible not merely for recognizing unity and order in the sensible manifold, but also for establishing order of the sensible manifold in the first place. As Hannah Ginsborg illustrates this point:

> Recognizing this as a dog implies recognizing that I ought to synthesize my representations in one way rather than other, for example, that I ought to see the tail as belonging with the head and legs rather than with the tree in the background, or that I ought to reproduce prior perceptions of barking, rather than, say, mewing or neighing. Recognizing the applicability of a concept, then, is recognizing a normative rule which governs the activity of my imagination in its reproduction of the manifold. It is because concepts serve in the first instance to specify ways, in which the manifold ought to be synthesized, not just ways in which the manifold is synthesized, that they can be identified with rules for the synthesis of the manifold.
>
> (1997: 51)

Concepts prescribe how the imaginative synthesis of the sensible manifold should be carried out and how the discrimination among the sense intuitions should proceed. They determine how we will come to perceive the object.[3]

In the third *Critique*, however, Kant introduces a broader form of cognition to which he refers to as the cognition in general (Erkenntnis überhaupt). It is the conditions of this general form of cognition that free harmony appears to satisfy, as it is evident from the following two passages:

> The subjective universal communicability of the kind of representation in a judgment of taste, since it is supposed to occur without presupposing a determinate concept, can be nothing other than the state of mind in the free play of the imagination and the understanding (so far as they agree with each other as is requisite for a cognition in general): for we are conscious that this subjective relation suited to cognition in general must be valid for everyone and consequently universally communicable, just as any determinate cognition is, which still always rests on that relation as its subjective condition.
>
> (KU 5:218; 103)

> The animation, of both faculties (the imagination and the understanding) to an activity that is indeterminate but yet, through the stimulus of the given representation, in unison, namely that which belongs to a cognition in general, is the sensation whose universal communicability is postulated by the judgment of taste.
>
> (KU 5:219; 104)

As Kant states, what belongs to cognition in general is not the mere play between cognitive powers of imagination and understanding (which can either be guided by concepts or free of them), but rather 'so far as they agree with each other' or are 'in unison'. He thus appears to identify cognition in general with the mental state of harmony between the imagination and understanding, irrespective of whether this harmony is attained by concepts (cognitive judgements) or it is in free play (judgements of taste). Yet, if free harmony belongs to the cognition in general and given that Kant identifies the awareness of free harmony with the feeling of pleasure – as he writes: 'no other consciousness of it [of a relation that is not grounded in any concept] is possible except through sensation of the effect that consists in the facilitated play of both powers of the mind (imagination and understanding), enlivened through mutual agreement' (KU 5:219; 104) – it follows that the feeling of pleasure represents a kind of cognitive awareness, namely, an awareness of cognition in general.

Unfortunately, Kant does not provide any further explanation as to what exactly cognition in general refers to. However, considering that it can be attained by free harmony, it is reasonable to assume that it refers to a type of cognition that is not necessarily determined by concepts. Kant indeed distinguishes such a type of cognition in his explanation of the empirical concept acquisition, as put forward in the Introduction to the third *Critique*. Presumably, the imagination has the ability to synthesize the manifold of intuition without being governed by the empirical concept, yet nevertheless exhibiting order and unity as required by the understanding. It is through such an act of imaginative synthesis, exercised in (logical) reflective judgements that we come to acquire empirical concepts in the first place.

Kant introduces a reflective judgement (i.e. a judgement that looks for the concept under which to subsume the sensible manifold) in order to account for our experience of the multitude of empirical properties that are left undetermined by the pure concepts of the understanding (a priori categories). He writes that even though we are in possession of pure concepts of the understanding, which determine nature in the most general way, they leave undetermined the empirical content of specific forms, such as dogs, stones, flower or particular events, such as the warmness of the stone being caused by the sun (FI 20:203; 9). Since these general, pure concepts of understanding do not determine the content of specific empirical forms, then without any further presupposition, there could be such a heterogeneity of empirical forms that we could never grasp and understand empirical world as a unified whole. There could be so many different ways of arranging these particular experiences that without the presupposition of some underlying unity we could never understand empirical world as a systematically organized whole (FI 20:209; 13). However, given that we do experience empirical forms in a purposive way and have a grasp of systematic relations that obtain among such empirical forms and laws (for example, a classification of biological forms into the system of genera and species), this indicates that in addition to general, a priori concepts of the understanding there must be some sort of principle that guides us in making our experience of the empirical world coherent and systematic. Kant identifies this principle of the reflective judgement as the a priori principle of the purposiveness or systematicity of nature (KU 5:183–184; 70):

> This principle can be none other than that of the suitability for the capacity of the power of judgment itself for finding in the immeasurable multiplicity of things in accordance with possible empirical laws sufficient kinship among them to enable them to be brought under empirical concepts (classes) and these in turn under more general laws (higher genera) and thus for an empirical system of nature to be reached.
>
> (F1 20:215; 19)

It is Kant's reasoning that the principle operates under the assumption that the empirical world forms a system in which all phenomena relate to each other and are divided into the genera and species, for it is this assumption that allows reflective judgements to look for the commonalities between empirical form and bring them under the universals. In other words, the principle of purposiveness presupposes a certain idea about the empirical world, namely, that it is as though it were organized by an understanding that is similar to ours, so that harmony is possible between our cognitive abilities and the empirical world itself. That is to say, the principle attributes a 'hypothetical purposiveness' to the empirical world (Fricke 1990: 47) in the sense that we perceive the empirical world as purposive (as if a product of an intentional activity), but not assuming that it is actually product of a human activity. Since empirical world is not actually constituted by the understanding, when in fact it does agree with it, such agreement is recognized as contingent. It is suggested by Kant that the principle is necessary for us to have empirical cognition. Only so far as we ground our reflection on the empirical world on the a priori principle of purposiveness of nature, namely, the assumption that 'nature has observed a certain economy suitable to our power of judgment and a uniformity that we can grasp' (FI 20:213; 17), can we 'make progress in experience and acquire cognition by the use of our understanding' (KU 5:186; 73). Furthermore, Kant writes that the principle does not produce any determinate cognition of the object, but only 'represents the unique way in which we must proceed in reflection on the objects of nature with the aim of a thoroughly interconnected experience, consequently it is a subjective principle (maxim) of the power of judgment' (KU 5:184; 71). The principle does not determine anything about the empirical world, but only represents an orientation we must take in our reflection of the world. The principle determines us and our need to perceive the world in a specific way – in a way that allows our understanding to cognize it. Thus, there is no quarantine that the empirical world will actually show its agreement with the principle.

In sum, Kant identifies cognition in general with our ability to cognize empirical world as a system and to attain unification of all our experiences.[4] It is within the purview of the reflective judgement, in contrast to determinate cognition, which belongs to the domain of determining judgements. Determinate cognition is necessitated by subsuming the manifold of intuition under the concept of the object, whereas cognition in general does not necessarily subsume under the concept, but rather under the general idea of the empirical world as a systematic whole, i.e. the a priori principle of the purposiveness. Accordingly, it is this principle that in the absence of determinate concepts of the understanding serves as a guiding rule for the synthesis of the sensible manifold. What remains to be seen is whether and in what way judgements of taste involve this principle.

Feeling of pleasure as the rule of structure

Kant considers the a priori principle of purposiveness mainly in its relation to the empirical concept acquisition, that is to say, as it operates in logical reflective judgements, where it serves as a guiding rule to find commonalities between different empirical forms. The principle is used here to find logical purposiveness in the relation between empirical forms (FI 20:216; 19). However, he suggests that the same principle is also responsible for finding aesthetic purposiveness, that is, purposiveness in the form itself, which leads to making a judgement of taste. As he writes, 'it is always possible and permissible, if experience shows us purposive forms in its products, for us to ascribe this to the same ground [i.e. principle of purposiveness] as that on which the first [i.e. logical purposiveness] may rest' (FI 20:218; 21). This idea comes out even more explicitly in the following passages:

> In a critique of the power of judgment the part that contains the aesthetic power of judgment is essential, since this alone contains a principle that the power of judgment lays at the basis of its reflection on nature entirely a priori, namely that of a formal purposiveness of nature in accordance with its particular (empirical) laws for our faculty of cognition, without which the understanding could not find itself in it.
>
> (KU 5:193: 79)

> Although aesthetic judgments themselves are not possible a priori, nevertheless a priori principles are given in the necessary idea of an experience, as a system, which contain the concept of a formal purposiveness of nature for our power of judgment, and from which the possibility of aesthetic judgments of reflection, as such, which are grounded on a priori principles, is illuminated a priori.
>
> (FI 20:233; 34)

The main idea suggested by these passages is that judgements of taste depend on an a priori principle, this principle being nothing else but the necessary presupposition of the purposiveness of nature that grounds our ability to find empirical concepts for the particulars. Accordingly, both finding the concept under which to subsume the particular sensible manifold and finding an object beautiful are made in reference to the same principle of purposiveness, and to the same cognitive need we have, that is, to systematize our experience of the empirical world. They both represent the satisfaction of the same principle of nature's purposiveness for our cognitive abilities.

The connection between judgements of taste and the a priori principle of purposiveness that guides our reflection on the empirical world and makes empirical cognition possible has been pointed out by numerous Kant's scholars (Horstmann 1989, Ginsborg 1990, Baz 2005, Matthews 2010: 63–79, Merrit 2014, Küplen 2015: 76–92, Palmer 2011).[5] In short, the idea is that beautiful objects represent the satisfaction of the same a priori principle of purposiveness that underlies our capacity to find empirical concepts for the particular. The difference is that in empirical concept acquisition, the principle is satisfied by means of finding a determinate concept, this latter referring to the general properties that different objects share with each other. According to Kant, a concept 'refers to the object indirectly, by means of a characteristic that may be common to several things' (A320/B377). This implies that reflection can result in a determinate concept only when we compare different empirical forms with each other in order to find general features among them, since only such general features can be explicitly communicated. However, in judgements of taste we reflect on a particular form itself, without comparing this form with others in order to find commonalities. As Kant writes, 'the ground of the pleasure is placed merely in the form of the object' (KU 5:190; 76). The object is judged as beautiful in virtue of its individual and distinctive properties that are directly available to our perception and which are more difficult or impossible to explicitly articulate.[6] Aesthetic content, as pointed out by many contemporary writers, is descriptively ineffable for it consists in the fine-grainedness of perceptual and expressive properties in contrast to the course-grained mental schemas that are necessary for identification and recognition of these properties (Davies 1994, Raffman 1993, Spackman 2012, Jones 2017). As Stephen Davies expresses this idea, our direct perceptual experience has a 'seamless plenitude that linguistic description does not retain' (1994: 159). Similarly argues Diana Raffman, who claims that while we are able to directly perceive and detect fine-grained properties, such as certain nuances of pitch and intervals or certain shade of colour, we are unable to remember them and consequently to identify them by means of verbal expressions:

> Although in hearing the nuances you know how they sound, there will be nothing you can say, no report you can make, that could by itself serve to reproduce that knowledge in the mind-brain of another listener. On the contrary, your only hope of getting him to know what you know is to ostend the signal; he must hear it for himself. To put the point another way: you must show him, you cannot tell him, what you know.
>
> (1993: 88)

It is due to the limitation of our cognitive system (in fact it is precisely because of its limitation that we form schemas and categories for easier recognition) that we are unable to process fine-grained perceptual nuances afforded by our sensory system. Aesthetic content can only be distinguished by a direct observation and direct experience of the work.

Now, since these particular, fine-grained aesthetic properties cannot be explicitly articulated (they can only be distinguished by direct perceptual observation), it also follows that aesthetic reflection on such distinctive and particular properties of the object cannot result in a determinate concept. As pointed out previously, reflection can lead to a determinate concept only when used to find commonalities between different forms since only general features can be explicitly articulated. Hence, purposiveness of an individual form can be recognized through the feeling of pleasure alone. Finding an object beautiful, similarly to acquiring a determinate concept for the particular, reveals that the object agrees or fits with the a priori principle of the purposiveness that guides our perception of the world, the difference being that a beautiful object reveals the satisfaction of the principle at the most particular and concrete level which for this reason cannot be grasped in a determinate concept, but can be manifested in the feeling of pleasure alone. Thus, even though experience of the beautiful does not result in a determinate concept, it does nevertheless satisfy the need of our reflective power of judgement to find purposiveness in the empirical world, that is, to find the agreement between the empirical world and our cognitive abilities.[7] As Anthony Savile suggests similarly, a beautiful object 'appears to cater for a need that we have to make cognitive sense of the world' (1993: 89).

It follows from this discussion that the role of the feeling of pleasure in judgements of taste appears to be similar to that of a determinate concept in cognitive judgements.[8] The feeling of pleasure is the way one recognizes purposiveness or free harmony in an individual object, just as a determinate concept is the way one recognizes the purposiveness (harmony) of an object's general properties. The feeling of pleasure in judgements of taste substitutes for the role of determinate concepts in cognitive judgements. Kant alludes to such an idea when he writes: 'As if beauty were a property of the object and the judgment logical (constituting a cognition of the object through concepts of it), although it is only aesthetic' (KU 5:211; 97).

We can pursue the idea of the similarity between the feeling of pleasure and a concept even further and show that the feeling of pleasure plays not merely a recognitional role, that is, being an awareness of the free harmony between the faculties of imagination and understanding, but also an organizational role – as a provider of order and unity to the sensible manifold. To recall, Kant characterizes concepts responsible not only for recognizing unity and order in the sensible manifold, but also for establishing that order in the first place. Now, while the idea of the feeling of pleasure serving as an alternative to determinate concepts is often dismissed as 'very unlikely' (Rogerson 1998: 123), I will argue in what follows for the opposite view. Specifically, I will attempt to show that the relationship between the feeling of pleasure and free harmony between cognitive powers of imagination and understanding is analogous to the relationship existing between determinate concepts and cognitive (conceptual) harmony in that the feeling of pleasure functions not merely as a means by which we come to recognize a relationship between imagination and understanding as freely harmonious, but also as a felt sense that guides us in ordering the sensible manifold and making free harmony possible. Let me illustrate this idea by means of Sigalit Landau's contemporary video artwork *Dead Sea* (2005).

Landau's work features hundreds of watermelons floating on the Dead Sea, joined together by a string forming a circle. Between the watermelons, some of which are open thereby revealing the intense red colour of their flesh, lies the artist's naked body. One of her arms is placed by her side, while the other one is stretched out, touching the open flesh of a watermelon. The video shows, in slow motion, how the string is pulled, thereby rotating the artist's body along with it until the circle is completely untied and out of sight. The video affords a stunning visual experience. However, there is much more to this work than its visual images being pleasing to the eye. Namely, these images work as aesthetic attributes that bring to mind various introspective, affective and emotional aspects that constitute the semantic content of abstract concepts such as the concept of the inevitability of death or the idea of inseparability of life, pain and struggle. For example, each image of a watermelon functions as an aesthetic attribute standing for a year in one's life. Watermelons are pulled by an unknown source, the image of which stands as an aesthetic attribute for the idea of powerlessness and determinism. The naked body of the artist, pulled along by the string, brings to mind the feeling of vulnerability and helplessness. Open watermelons, revealing the intense red colour of their flesh, are like open wounds, embodying the presence of pain and struggle in one's life. Watermelons are half submerged in the sea which may be an aesthetic attribute standing for the idea of life itself. Moreover, since it is the salt-saturated water of the Dead Sea, this in addition stands as an aesthetic attribute for the idea of an unavoidable exposure to the harsh realities of life. The collection of these aesthetic attributes inspires a multitude of thoughts, feelings, emotions, beliefs and other mental aspects that are involved in our reflection on the idea of the inevitability of death and its inseparability from life.

But how do we recognize Landau's work exhibiting these ideas? As pointed out previously, even though abstract ideas such as death, pain, life and struggle can be experienced (we can experience their concrete instances), they nevertheless lack a determinate empirical counterpart. They do not have a clear physical referent that can be sensory experienced. For example, the concept of the inevitability of death does not have unique physical features such as colour, size, form or texture that one could visually perceive. Such ideas refer to mental and emotional states and involve experiential features and other introspective properties, such as beliefs, desires, values, memories, intentions and goals. But these experiential and introspective features cannot be directly exhibited in ordinary experience. There is thus an indeterminate material to these concepts for which no sensible intuition can be given.

But if there is no direct sensible intuition given to these concepts, then this means that there is also no determinate way to demonstrate why the idea of the inevitability of death and Landau's work agree with each other. This is different in the case of a determinate concept such as the concept of a flower and its sensible intuition, the image of a flower, since in this case we can demonstrate clearly for why they are in agreement by simply pointing out some of its features, such as having leaves, petals and stem in a certain combination. While we can explicitly articulate criteria for why we would classify something as a flower, we cannot state such criteria that would identify a particular image as representing the idea of the inevitability of death. For example, we cannot explicitly point out as to why the aesthetic idea stimulated by the image of watermelons is in harmony with the idea of the inevitability of death. Nonetheless, we can still recognize that they are in harmony, the difference being only that this harmony is recognized through the feeling of pleasure.

It is the feeling of pleasure that guides us (both the artist and the viewer) in determining which collection of images and thoughts (that is, aesthetic attributes) is appropriate for the expression of an idea. Grasping the meaning of the work is a slow process of testing which or what combination of aesthetic attributes feels suitable for capturing the concluding idea. Many different images and thoughts might come up during this interpretative process, but not all of them resonate well with each other. As Kant writes:

> To give this form to the product of beautiful art, however, requires merely taste, to which the artist, after he has practiced and corrected it by means of various examples of art or nature, holds up his work, and after many, often laborious attempts to satisfy it, finds the form that contents him; hence this is not as it were a matter of inspiration or a free swing of the mental powers, but a slow

and indeed painstaking improvement, in order to let it become adequate to the thought and yet not detrimental to the freedom in the play of the mental powers.
(KU 5:312; 191)

The ability to find the appropriate form for the expression of an idea is a matter of taste, which provides intelligibility and meaning to the multitude of aesthetic attributes or associational thoughts. Taste or the feeling of pleasure, as the manifestation of the principle of purposiveness, has an orientational function; it guides us towards structuring the manifold of aesthetic attributes in one way rather than the other. For example, that I ought to associate the image of open watermelons with wounds and pain rather than with something else, say, the summer time of my youth as that brings more coherence and sense to the manifold of attributes. In the process of discriminating between appropriate and inappropriate forms we are guided by the pleasure of taste. Pleasure functions not merely as an awareness that a fit between the free imagination (that is, the combination of aesthetic attributes that is not governed by determinate concepts) and understanding's need for unity and order has been found, but also as a felt sense that guides the imagination to synthesize aesthetic attributes in a certain way, namely, in a way that maximizes coherent meaning and understanding of the concluding idea. We recognize the appropriate ordering of aesthetic attributes by the sense of satisfaction it produces in us, whereby we come to feel that all the features of the manifold are brought together and make sense. And the opposite is also the case, inappropriate ordering of the manifold of aesthetic attributes provokes the feeling of dissatisfaction and tension as it is less sense-making. The feeling of pleasure (or displeasure) signifies the compatibility (or incompatibility) between the form and thought, respectively.

Affinity between the feeling of pleasure and determinate concepts

The idea of the feeling of pleasure playing a role in judgements of taste that is similar to the role determinate concepts play in cognitive judgements is not formally developed by Kant. There are, however, many passages throughout the third *Critique* that can be read in a manner of supporting this idea. For instance, Kant describes the a priori principle of purposiveness that guides logical and aesthetic reflection as 'fundamentally identical with the feeling of pleasure' (FI 20:230; 31–2), suggesting thereby that the principle guides our reflection through the feeling of pleasure.[9] Presumably, the feeling of pleasure has an inherent judging capacity, which is furthermore evident from Kant's definition of taste as 'the faculty for judging through such a pleasure' (KU 5:190; 76). Similarly, he writes that the feeling of aesthetic pleasure or displeasure 'grounds an entirely special faculty for discriminating and judging' (KU 5:204; 90). Statements such as these give support to the idea that the feeling of pleasure has an inherent capacity to discriminate fine aesthetic differences among empirical objects and to conduct the activity of imagination and understanding into a free harmony. But this is the role that concepts play in cognitive judgements, that is, they govern the agreement or harmony between cognitive faculties of imagination and understanding.

Likewise, Kant frequently claims that the feeling of pleasure plays a role of a predicate in aesthetic judgements of taste and thus performs a function similar to that of a determinate concept in cognitive judgements. For instance, he writes that 'all judgments of taste are also singular judgments, since they combine their predicate of satisfaction not with a concept but with a given singular empirical representation' (KU 5:289; 169). He suggests similarly in the following: 'A perception can also be immediately combined with a feeling of pleasure (or displeasure) and a satisfaction that accompanies the representation of the object and serves it instead of a predicate, and an aesthetic judgment, which is not a cognitive judgment, can thus arise' (KU 5:288; 168). Predication refers to an act of judgement whereby a singular representation is subsumed under a general representation, this being a determinate concept in ordinary cognitive judgements. The concept serves as a predicate in the sense that it 'specifies the logical form which determines the wholeness or order of the various parts of the sensible manifold' (Tinguely 2013: 224) and thus serves as a rule determining how the sensible manifold should be organized together to attain the agreement between cognitive faculties. If the feeling of pleasure also has a predicative function, as Kant appears to suggest in the mentioned passages, then this means that it can also serve as a rule for the unification of the manifold. The difference is that feeling of pleasure serves as a predicate providing an additional unity that comes up subsequently to the logical unity afforded by the concept of the object. As Kant states: 'That judgments of taste are synthetic is readily seen, because they go beyond the concept and even the intuition of the object, and add to that as a predicate something that is not even cognition at all, namely the feeling of pleasure (or displeasure)' (KU 5:288; 169).

One way to account for this additional aesthetic unity is to claim that the feeling of pleasure serves as a rule for the synthesis of those specific and distinctive features of an object that are left undetermined by the concept of the object. I mentioned previously that according to Kant's epistemological theory, a determinative judgement is necessary in order to have perceptual experience of an object. The form of the object (combination of sensible manifold) is determined by the concept to some degree. In order to recognize a particular object, say a dog, the imagination must follow the dog-rule and combine specific features such as tail, leg, head, body and fur as the dog-rule prescribes. Without this cooperation between the imagination and understanding there would be no perceptual experience of an object.

Yet, a particular image of a dog has a distinct combination of its general features. But these distinctive features of a particular dog are not entailed by the concept of a dog, and the object of aesthetic reflection is always a particular determination of the sensible manifold. Accordingly, even though we have a concept of a dog in virtue of which we come to recognize a particular sensible manifold as a dog, this concept does not determine our aesthetic reflection, simply because our reflection takes into account those distinctive and individual properties that are not entailed by the concept of a dog. Even if one possesses a more specific conceptual vocabulary, such as the concept of a Chihuahua, this concept is still insufficient to fully determine the particular sensible manifold. The concept of a Chihuahua still represents general features that all Chihuahuas have in common and which can be thus explicitly communicated. For example, a Chihuahua is a dog distinctive by its small size, large and erect ears, round eyes well set apart, and well-rounded, apple-like head shape. One can recognize a particular dog as a Chihuahua by recognizing these general features in the particular sensible manifold. However, these general features are still inadequate to fully determine all the distinctive details of a particular Chihuahua as it is presented to our direct perceptual experience. The concept can be well particular and determinate (Chihuahua instead of a dog); yet, it always represents an object in virtue of general properties shared with others of its kind and which can be explicitly articulated. As such it can never fully determine the particular sensible manifold that is directly available to our perceptual observation. It is the presence of these additional, conceptually undetermined features that aesthetic reflection takes into account. The feeling of pleasure serves as a rule for the unification of this additional feature that is left undetermined by the concept of the object. It is the feeling of pleasure itself that selects which conceptually undetermined properties will be picked out, made salient and gathered together into a unified form. Feeling of pleasure organizes the sensible manifold to support the free harmony.

Although the interpretation of the feeling of pleasure as playing a role of a concept in judgements of taste is not widely acknowledged among Kant's scholars, the central elements of this position can nonetheless be found in the works of numerous writers. For example, Jean-François Lyotard identifies judgements of taste with the feeling of pleasure and claims that the latter serves as a 'guide for thinking' (1994: 26). Similar suggestion is made by Richard Aquila, who argues that the feeling of pleasure plays an active role within the workings of our faculties of imagination and understanding. As he states, 'a pleasure is a pleasure in the form of some object not by regarding it as a mere consequence of certain activities involved in apprehending that form, but rather by regarding it as itself "informed" by the very same formal structures as are involved in the perception in question' (1979: 27). Joseph Cannon emphasizes the active role of aesthetic pleasure in judgements of taste by claiming that it contributes to the animation of our cognitive faculties. He writes that 'judgments of taste must be capable of being simultaneous with acts guided by them' (2008: 60). Without such conception of pleasure we would have difficulty explaining the productive activity of the artist which consists primarily in their ability to 'track changes in his or her presentation as approaching or failing to approach beauty' (2008: 60) That is to say, aesthetic feeling of pleasure plays an active role not merely in the editing phase of the artist's creation, namely, in judging whether the product is beautiful or not, but also during the artist's creative process itself, where it is manifested in 'seeking the movements or gestures (physical, rhetorical, or both) that produce, or are, a beautiful presentation' (2008: 60). The feeling of pleasure is an essential part of the productive practice of art as it directs and guides the artistic activity towards achieving beauty (free harmony).

It is, however, Joseph Tinguely that most openly endorses the reading of aesthetic pleasure as having a close affinity with the determinate concept. According to his position, called internal objectivism, the feeling of pleasure is neither a prior occurrence nor the consequence of the judgement of taste, but rather an internal element of the judgement itself 'somehow giving it shape and influencing its outcome – as though feeling could itself be a discriminatory capacity' (2013: 213). He proposes the possibility of the feeling of pleasure having the ability to synthesize the sensible manifold, that is, of 'picking out and holding together aesthetic attributes so that absent the pleasurable feeling a judging subject could not (or would not) attend to the object or scene in the manner by which its aesthetic properties become salient' (2013: 219). Presumably, such an account, while not necessarily intended by Kant, provides a completion to the problem of the relationship between concepts and intuition that Kant develops in the first Critique:

> It is precisely the nuances of this relation which come under careful consideration in the instances of 'free beauty.'[...] In the case in which experiences of 'free'

beauty are not to be read as falling back on a moment of a purely perceptual 'given', the key question is, in the absence of determining concepts, what else could possibly do the work of actively synthesizing or schematizing the manifolds of sense? [...] Kant must allow that a feeling of pleasure could be active in the process by which various parts of the empirical manifold are seen to 'hang together' to form an ordered, unified whole.

(2013: 223)

One can indeed find support for such an interpretation in Kant's claim that 'since the freedom of the imagination consists precisely in the fact that it schematizes without a concept' this means that 'the judgment of taste must rest on a mere sensation of the reciprocally animating imagination in its freedom and the understanding with its lawfulness, thus on a feeling that allows the object to be judged' (KU 5:287; 167). Here the suggestion is that in the absence of concepts it is the mere feeling of the activity between cognitive powers that performs the synthesis of the manifold. Moreover, considering Kant's statement that the relationship between the imagination and understanding can be sensed by means of the feeling – as he writes 'one can also consider this relation of two faculties of cognition merely subjectively, insofar as one helps or hinders the other in the very same representation and thereby affects the state of mind, and [is] therefore a relation which is sensitive' (FI 20:223; 25) – this allows for the possibility that the feeling of pleasure itself has the ability to affect the direction of that relationship. The feeling of pleasure can lead the activity between cognitive powers into free harmony. In other words, the feeling of pleasure can serve as the guiding rule as to how to synthesize and combine together the sensible manifold to achieve harmony and thus it can play the role of the concept.

Aesthetic reflection and objectual understanding

I argued thus far that the function of taste is to provide sense and meaning to the sensible manifold; hence, it must serve as a kind of cognition. While judgements of taste are non-cognitive judgements in the determinate sense of cognition as the ability to apply concepts to the sensible manifold, they nevertheless have a cognitive function in the general sense of cognition as the systematic ordering of experiences into meaningful wholes. Taste is the manifestation of the principle of purposiveness that guides our organization of the empirical world through the feeling of pleasure (or displeasure) and thus must be seen as a form of cognition. Specifically, taste serves as a rule for the organization of the additional, conceptually undetermined aesthetic material, which cannot result in a concept determined by words, but rather in the feeling of pleasure alone. Kant hints at the possibility of the existence of the beautiful as a kind of cognition when he writes: 'That is beautiful which is cognized without a concept as the object of a necessary satisfaction' (KU 5:240; 124). In other words, the beautiful is not something that we simply feel through pleasure; rather, we cognize it through the feeling of pleasure, the implication being that the mind can cognize through feelings; the mind can feel (or sense) cognition. The aesthetic feeling of pleasure in the beautiful can carry forward some form of cognitive insight.[10]

Even though judgements of taste do not result in a determinate concept and hence do not, as Rudolf Makkreel states, 'add to our stock of empirical knowledge' (1990: 51), they can nevertheless contribute to our cognition of the world. As pointed out by contemporary epistemologists, not all our cognitive experiences necessarily result in a determinate linguistic expression. The epistemic concept of objectual understanding is a case in point. Cognition results in a determinate concept when reflecting on the common features of the object, since only common feature can be explicitly articulated. Yet, the subject of aesthetic reflection is the particular object itself in all its structural details, which cannot be explicitly described. For example, according to Gottfried Gabriel, aesthetic reflection on the particular object gives rise to a type of aesthetic cognition characterized as 'determinate indeterminacy'. As he writes: 'The particular in this sense is determinate insofar as it is a singular entity, but indeterminate with respect to its meaningfulness' for it evokes inexhaustible meanings and cognitions, and thus is 'fundamentally open to a completely different interpretation or view of things' (2013: 187). Bennet Reimer makes a similar observation:

> Some of these mental structurings can lead to conceptualization because they deal with common features. But it is likely that many others are not of this sort at all in that they are responses to particular instances which are themselves deeply meaningful (...) Such an experience is "vertical" in its affect: it is rich with knowing as a singular presence rather than as a horizontal commonality with similar things. The knowing or awareness or import or comprehension in such an experience is nonconceptual.
>
> (1986: 113–14)

A. W. Moore concurs as he identifies the experience of free harmony between imagination and understanding with the 'feeling of unity' and describes it as a type of inexpressible knowledge. This is the kind of knowledge that' – this is a normal sentence that I write and thus should not be intended as a citation. There is citation after that, starting "lacks content of the sort that can be expressed by a declarative, descriptive sentence – and that such knowledge, some of which enables whoever possesses it to make sense of things in a certain way, is required to ground any fully conceptual expressible knowledge." (2007: 477)

One of the characteristics of aesthetic form of cognition, besides being inexpressible, is that it involves freedom from our projection of determinate meanings into the object. Aesthetic reflection is not concerned with finding commonalities among objects and thus with delivering a particular determinate cognition. In contrast to conceptually governed harmony between cognitive powers in ordinary cognition, free harmony is not determined by a particular concept and therefore it is not restricted by seeing and thinking of an object in a particular way. Aesthetic appraisal, as Henry Allison writes, 'involves a suspension of our ordinary cognitive concern with classification and explanation' (2001: 187). The free harmony between imagination and understanding signifies a freedom from determinate perception and thinking. Both understanding and imagination are free in their mutual endeavour – the former is free for it is not concerned with the identification of the object and the latter enjoys freedom as it is not governed by any particular empirical concept. Since aesthetic reflection is not limited by the conceptual determination and thus by considering the object from well-defined perceptual and semantic viewpoints, it can be fully engaged with reflecting on the mere structure of the object. We cannot be fully aware of the structure and pattern of the object, as well as of its sensory qualities such as colours, textures, shapes and sounds, as long as we perceive the object in light of conceptual determination, that is, as long as we perceive these qualities 'submerged in the object' (Dorter 1990: 39). The reduction of conceptual influence on a particular object gives us the possibility to detect, discover and explore features of the object, as well as its various relationships that have been left unnoticed so far. Kant describes such non-conceptual aesthetic reflection as disinterested – a way of seeing the object in which we are indifferent to its existence:

> If the question is whether something is beautiful, one does not want to know whether there is anything that is or that could be at stake, for us or for someone else, in the existence of the thing, but rather how we judge it in mere contemplation [...] One only wants to know whether the mere representation of the object is accompanied with satisfaction in me, however indifferent I might be with regard to the existence of the object of this representation.
>
> (KU 5:204–205; 90–1)

Aesthetic reflection is not restricted by the question of the representation (that is, what the object should be) and thus with the practical significance of the object in its everyday existence, but rather how it merely appears to us as we directly observe its formal qualities – the finer distinctions of forms, lines, texture and colours. This opens new ways of seeing the object and new ways of making sense of it. Aesthetic reflection gives us the opportunity to discover new meanings that the object can have for us.[11]

As pointed out previously, aesthetic reflection relates the object not to the concept of the object for the sake of determinate cognition, but rather to the subject and its feeling of pleasure or displeasure. Thus, while aesthetic reflection cannot produce any change in our ordinary cognition of the object in the sense of ascribing any objective properties to the object, it can nevertheless produce a change in us as to how we come to see and interpret the object. That is, aesthetic reflection discloses not what the object is, but rather our relation to it and what it means to us. I believe this is the idea Kant has in mind when he writes that artistic beauty as an expression of an aesthetic idea enlarges the concept, not logically, but rather aesthetically in an indeterminate way, by bringing to mind a multitude of thoughts, feelings, moods and sensations (that is, our subjective experiences) connected with the given concept. In other words, expression of an aesthetic idea enhances and broadens the experiential dimension of our concepts. Our experience of the artistic expression of aesthetic ideas reflects our own perception and attitudes of the world – our own thoughts, feelings, emotions, beliefs associated with abstract concepts and rational ideas, yet, in a comprehensive and coherent way. In other words, artistic expression of aesthetic ideas embodies a form of (objectual) understanding that has as its basis the grasp of the explanatory relationship between different experiential properties that constitute the content of abstract phenomena. Feeling of pleasure in the experience of aesthetic ideas represents the attainment of our understanding of the world of ideas and concepts that have no empirical counterpart, and for which no evidential support in the empirical world can be found. This understanding does not determine anything about the empirical world, that is, it does not make any claims about the world in the sense of ascribing objective properties to it, but rather it determines us and our subjective engagement with the world, which is governed by our cognitive need to make coherent sense of things. As Rudolf Makkreel makes this point: 'Aesthetic judgments also refer to objects, but disclose more about our state of mind than about what is apprehended' (2006: 233). Consider again Sigalit Landau's artwork *Dead Sea* (2005). We can notice that the work does not ascribe any objective properties to the idea of the inevitability of death or the idea of inseparability of life and death, but rather it makes salient various thoughts, feelings, moods and other mental representations associated with the idea we have of the inevitability of death. For instance, how it is determined by the implicit feeling of powerlessness as occasioned by the sight of the artist's naked and exposed body silently slithering through the salt-water while locked into a slowly unravelling spiral of watermelons, or by the idea of vulnerability and pain as conveyed by the image of the artist touching the 'wounded' watermelons that reveal the intense red colour of their flesh, while at the same time being susceptible to the sting of salt-water. By revealing the explanatory relations of various experience-related properties underlying the idea of the inevitability of death, the work also gives us the opportunity to grasp the implications and significance of this idea for our own lives. For example, that life is a form of a survival against the thread of death that is always present to life and that it can never attain complete freedom from the traces of death because death is its other without which it cannot exist. But if the cause of the vulnerability and pain of life is the death itself inevitably existing within life, then the only way to live well is to acknowledge and accept death – to slide silently through the stingy water of the Dead Sea.

Conclusion

The main aim of this book was to reconcile the disagreement between aesthetic cognitivism, the view that works of art serve as an important source of knowledge about the world and that knowledge thus obtained contributes to the aesthetic value of an artworks, and aesthetic anti-cognitivism, which claims that artworks cannot serve as a source of non-trivial, unique and aesthetically relevant source of knowledge. Anti-cognitivists correctly point out that aesthetic value of artworks lies in the interrelation between work's formal structure and thematic exploration of its content, and thus mere cognitive claim deprived of the formal and contextual structure cannot have aesthetic relevance. Cognitive value of artworks has no bearings on their aesthetic value. Anti-cognitivists restrict the notion of cognitive value to the propositional knowledge and factual truths, and rightly conclude that the consideration of the truth value of thematic statements or ideas presented by the artwork is not part of our engagement with the work qua artwork, and that as soon as we engage in cognitive reflection on the truth value of thematic statements, we fail to appreciate the work in an aesthetic way. Aesthetic cognitivists, on the other hand, point out that cognition need not be restricted to mere propositional knowledge, but it can be employed in a broader way, including experiential knowledge, philosophical or conceptual knowledge and type of cognition that can be grasped, not in terms of knowledge, but rather in terms of axiological understanding. However, the problem with most of these cognitive approaches is that when applied to art they face difficulties accommodating the textual-constraint criteria of aesthetic cognitivism, which claims that to ascribe proper cognitive value to works of art, this value must be part of the work's content. Yet, given that the content of many works of art is fictional and therefore their conveyed cognitive claims cannot pertain to the real world, the usual move taken by aesthetic cognitivism is to highlight the employment of additional skills in order to extricate cognitive claims from the content – either to look for the cognitive value on the thematic level (propositional knowledge approach), to engage our imaginative skills of mentally simulating what it is like to be a certain fictional character and applying this knowledge to understand real-life people (experiential knowledge approach), or to utilize our reflective capacities to extract philosophically relevant conclusions from the fictional content (conceptual or philosophical approach). Each of these approaches thereby fails to accommodate the textual constraint argument as the cognitive value ascribed to works of art is not actually a property of the content of the work. Accordingly, aesthetic cognitivism appears to be facing the following dilemma: we search for the property of being cognitively valuable either in the work's content or outside of it. If we search for the cognitive property within the content, then the work cannot have this property since the content is fictional and does not have a real-world reference (paradox of fiction). But if we search for the cognitive property outside the work, then it is not the work itself that possesses this property (textual constraint argument). Hence, in either case, the artwork does not have the property of being cognitively valuable. Presumably, aesthetic cognitivism fails.

John Gibson's account proposes a way to avoid this dilemma by moving artwork's cognitive value from knowledge to the concept of understanding, for the latter does not demand factuality and can be thus obtained from the fictional content. However, while his proposal of distinguishing between knowledge and understanding can indeed indicate the way to proceed in formulating a positive account of aesthetic cognitivism, the particular form of understanding that he argues for, namely, axiological understanding, faces a difficulty of being articulated as a genuine cognitive achievement. Rather, as I argued, the objectual understanding, as a distinctive epistemic concept, constituted by the internal grasping of the various explanatory relationships in the body of information, can account better for the specific cognitive value offered by works of art. An exemplar of such objectual understanding in respect to art is given by Catherine Elgin, which I further developed in light of Kant's theory of art as an expression of aesthetic ideas. I argued that works of art promote (objectual) understanding of abstract phenomena in that they connect the indeterminate material, namely, the introspective, affective and emotional properties, which constitute the semantic content of abstract concepts and ideas, with particular imaginative representations (or aesthetic attributes) and thereby making salient the various relationships between such properties. I furthermore established the aesthetic relevance of objectual understanding thus achieved by claiming that the comprehension of various introspective, emotional and affective properties associated with given abstract concepts proceeds by means of the feeling of pleasure. It moves us aesthetically to make a coherent and compelling sense of the various experience-related properties that appear to be central to the content of abstract phenomena. Moreover, I demonstrated that the feeling of pleasure in artistic expression of aesthetic ideas serves a role not merely of recognizing the attainment of (objectual) understanding, but also as a felt sense that guides us in ordering different elements into a coherent and systematic whole, thereby making objectual understanding itself possible. I argued that the feeling of pleasure is a manifestation of Kant's principle of purposiveness, employed in reflective judgements, that guides us in ordering the aesthetic material one way rather than another, in a way that brings coherence and understanding of the various semantic material. Each artistic expression of an aesthetic idea produces a different understanding of abstract phenomena depending on the artist's own explanatory perspective that guides their choice of particular aesthetic attributes (imaginative associational thoughts) in order to highlight certain features in terms of causal relationships, thereby producing different interpretations of meanings that abstract phenomena can have for us.

Such an account of aesthetic cognitivism can accordingly meet all the anti-cognitivists objections. Given that objectual understanding does not promote new truths, but rather have as its aim making sense of already-attained truths, particularly those truths that are important for our own well-being, its cognitive value is not trivial. Even though we know the logical meaning of abstract and rational concepts such as truth, alienation, hopelessness and immortality, we have difficulties understanding and making sense of them as they do not have direct empirical counterparts. They refer to mental and emotional states, situations and state of affairs and thus they are determined by properties that express experiential information, emotional aspects and other introspective properties, such as our own beliefs, personal memories, intentions and other personal-life concerns, and these properties cannot be directly exhibited in ordinary experience. To fully understand the meaning of such concepts requires comprehending how these various experience-related properties are related to each other in terms of causal and explanatory relationships, and their implications and significance for our own lives. Works of art, as expressions of aesthetic ideas, can provide such understanding as they are able to connect these properties with particular imaginative representations (or aesthetic attributes) and thereby to help us overcome cognitive limitations that we often experience in our attempts to articulate the meanings of various abstract phenomena. Furthermore, since this objectual understanding is carried forward by the content of the work, the aesthetic attributes or imaginary representations that bring to mind various thoughts, feelings, emotions and other mental representations that constitute the content of abstract phenomena, this account satisfies the textual constraint condition. In addition, given that cognitive effect of artworks is evaluated not in terms of their correspondence to the external world, but in terms of their cultivation of objectual understanding (grasping the coherence and explanatory making relationships), this approach can avoid the no-justification objection in terms of providing argumentative and evidential warrant. Objectual understanding involves the epistemic state of first-person reflective availability, namely, one's internal seeing or grasp of the relationships between different elements that constitute the meaning of abstract phenomena. One can tell from the inside, based on one's own conscious awareness, whether the epistemic state of understanding has been attained. Since, as I have argued, this awareness is given precisely through the feeling of pleasure, my account can therefore satisfy the aesthetic relevance condition of aesthetic cognitivism. The aesthetic feeling of pleasure is a powerful felt dimension of experience that functions importantly in what we perceive and think. In this respect, pleasure of taste has a cognitive function similar to the function that determinate concepts play in our ordinary cognition, namely, it serves not merely as a means of recognizing the attainment of understanding, but also as a felt sense that governs our organization of the various imaginative-associational thoughts (aesthetic attributes) and thereby provides meaning to it.

Kant's aesthetic theory accordingly provides us with a philosophical resource to reconcile the debate between aesthetic cognitivism and anti-cognitivism in respect to art for it allows us to interpret the aesthetic feeling of pleasure not as a mere passive awareness of the formal properties being in accordance with our aesthetic sense, as suggested by the traditional conception of aesthetic value, but rather as an active discriminatory power that guides us in arranging and holding together properties of an object in a way that brings coherence, significance and meaningful understanding of it. Aesthetic feeling of pleasure, as a determining basis of object's aesthetic value, has a discerning power of making sense of things and thus can be regarded as a form of cognitive value. Such an interpretation of aesthetic value can thereby secure a direct route to aesthetic cognitivism, for it allows us to consider aesthetic appreciation of artworks as a unique form of cognitive pursue, in particular, the kind of cognitive pursue that has as its aim an understanding of abstract phenomena as they are determined by our own subjective experiences. This significantly differs from the prevalent view of aesthetic cognitivism, which takes cognitive value as distinct from aesthetic value, yet that on certain occasion the former can bear on the latter. The integration of the cognitive into aesthetic presumably occurs when the cognitive insight, as embedded in the content of the work, is manifested in the aesthetic details of the work – its formal, stylistic and expressive properties. However, such an account, which interprets aesthetic value as merely integrated with the cognitive value, rather than being a form of cognitive value itself, is highly sensitive to anti-cognitivist's objections. According to aesthetic anti-cognitivism, the aesthetic value of an artwork consists in the appreciation of the formal and imaginative presentation of a theme, rather than in the theme itself and what it communicates. It is precisely this attention to artwork's mode of thematic presentation that presumably precludes any direct relationship between the artwork and its real-life epistemic benefits. Arguably, while we can consider works of art as a form of cognitive pursuit, mirroring the human world, they cannot at the same time function as a form of aesthetic pursuit, concerned with the perfection of the formal and imaginative qualities and aesthetic pleasure of the beautiful, for as soon as we regard works as concerned with the facts of the human world, we abandon our attention to the aesthetic dimension of the work (Lamarque and Olsen 1994: 408–9, Diffey 1995). That is to say, we cannot apply cognitive vocabulary as a standard to evaluate artwork's aesthetic success and vice versa. Such conclusion, however, is unwarranted if we take aesthetic appreciation itself as a unique type of cognitive appreciation, specifically as an (affective) appreciation of the internal awareness or comprehension of the formal relationship between different elements of the phenomena. The aesthetic feeling of pleasure is constitutive of our cognitive abilities in that it is a form of cognitive pleasure, namely, a pleasure at attaining the comprehension of the internal relationship between the body of (perceptual and semantic) information. Aesthetic appreciation, as an appreciation of the comprehensive structure of the object, has a necessary cognitive effect, this effect admitting of different degrees. The more comprehensive and coherent is the relationship between different elements of the phenomena, the higher is the degree of (objectual) understanding attained and consequently the higher its aesthetic value. Aesthetic and cognitive values are intrinsically intertwined and they necessarily determine each other. To aesthetically value a work of art entails valuing it for its capacity to obtain (objectual) understanding. This implies that cognitive value (understood in terms of objectual understanding) is essential to the aesthetic practice of art appreciation and not its mere by-product. Furthermore, it allows the possibility that works of art have other forms of cognitive value, such as imparting factual truths about the world, refining our conceptual vocabulary, promoting emotional and experiential knowledge, without these cognitive merits having any necessary effects on artwork's aesthetic value. Cognitive value of works of art, considered in the Kantian sense as expressions of aesthetic ideas, is aesthetically relevant insofar as it is grasped in terms of (objectual) understanding of abstract phenomena, for the latter requires for its attainment free harmony, that is, the aesthetic mode of arranging together various introspective, emotional and affective aspects that appear to be central to the semantic content of such phenomena.

Notes

Introduction

1. The notion of aesthetic value is used here in the broader sense, including not only narrow aesthetic properties such as formal properties, but also expressive and representational properties among others. In this respect, aesthetic value functions as an equivalent to artistic value.
2. Aesthetic cognitivism does not claim that all works of art have cognitive value, but only that some of them have, irrelevant whether they are representational or non-representational works. Moreover, not all artworks that have cognitive value are aesthetically better. Works of art can serve as a source of knowledge without this having to do anything with their aesthetic value. In addition, aesthetic cognitivism does not claim that cognitive value of artworks completely determines their aesthetic value, as they grant that there is a plurality of aesthetic/artistic values, cognitive value being one of them.
3. However, Antony Aumann (2014) defends the opposite direction of the cognitive and aesthetic relationship, namely, that aesthetic value determines work's cognitive value, specifically, philosophical value.
4. For a version of this argument see also Graham (2005: 59).
5. Similar argument is given by Young (2001: 108–9).

Chapter 1

1. Version of this argument has also been forwarded by Novitz (2004) and Gibson (2008).
2. For a version of this argument see also Carroll (2018: 22) and Phelan (2021: 156–7).
3. One of the earliest advocates of the view that artwork's cognitive value lies in offering hypotheses, rather than true beliefs, is Peter Mew (1973).
4. A similar version of propositional knowledge approach has been proposed by James Young (2001). According to his approach, works of art provide us with propositional beliefs, which are derived from the right perspective that the artwork presents, whereas the rightness of a perspective is illustratively demonstrated (by means of devices such as selection, amplification, simplification, juxtaposition,

correlation, etc.). Yet, to know that the perspective presented by the work is the right one, the audience must submit the perspective to empirical testing. That is, if the perspective 'accords with my experience' and 'helps me to make better sense of the phenomena than alternative perspectives', then this is the right perspective (2001: 106). But this means that it is not the work itself, but rather the audience that provides justification for the perspective presented by the work, and thus this proposal faces similar objection than Kivy's account. For a more detailed criticism of Young's account, see also Matheson and Kirchhoff (2003).

5   As originally argued for by Frank Jackson (1986).
6   I discuss such self-knowledge acquisition from works of art in Chapter 3.
7   Though, as interestingly pointed out by Davies (2010: 66), while the what-it-is-like experience might feel appropriate for the audience, one can never really be sure whether the affective experience reflects mere author's narrative skills or it is really representative of the depicted situation.
8   An exception is Wachowski Brothers' film, *The Matrix* (1999). The film satisfies the tripartite structure of a philosophical thought experiment in that it explicitly draws philosophical implications from the imaginary scenario, thereby revealing the general lesson. However, the film fails to meet the textual-constraint objection since it does not promote philosophical knowledge via the cinematic medium itself, that is, through particular visual narrative details.
9   For a more recent criticism of Gibson's account, see also Peels (2022).
10 For a distinction between knowledge and understanding, as well as differentiation between different kinds of understanding (propositional, explanatory and objectual), see Baumberger (2014).

Chapter 2

1   For a related reading of aesthetic ideas, see also Anthony Savile, who describes aesthetic ideas as 'concrete presentations of particular themes that are offered to us by individual works of art' (1987: 180). Robert Wicks explains them similarly, namely, as 'meaning-rich images' (2015: 26). In his recent work, Paul Guyer describes an aesthetic idea as an 'indeterminate sensory or imaginative representation' (2021: 622). Among others holding a similar interpretation are Allison (2001: 283–4), Sassen (2003: 173) and Crawford (1982: 156).
2   Samantha Matherne uses the term 'experience-oriented' aesthetic ideas for these kinds of concepts (2013: 21). She argues similarly that aesthetic ideas can represent not only moral and rational ideas, but also everyday kinds of ideas, concepts and feelings. Among others holding a similar view are Lüthe (1984), Rogerson (1986: 99) and Tuna (2016).

3   I believe Kenneth Rogerson has something similar in mind when he writes: 'While we may have some experience of such things their full import is yet beyond ordinary experience, for example our psychological attitudes to such things' (1986: 99).
4   According to Kant, a schema is a sort of an image and a rule at the same time, that is, a rule for linking a set of sense data with an appropriate concept (A141/B180). A schema represents an abstract image of the essential properties and the relations that obtain between them. For example, a schema of a flower contains the essential features of a flower, such as petals, leaves and stem in a certain combination. Even though there are different kinds of flowers, they all entail this rule in virtue of which they are recognized as flowers. Accordingly, a schema is necessary for us to have meaningful experiences.
5   The flower example pertains to the relation of concepts and particulars in general, whereas my investigation is deliberately restricted to the reception of artworks.
6   Among others making this point are Gorodeisky (2010) and Küplen (2015: 65–72).
7   My interpretation accordingly differs from the view that aesthetic attributes supervene on logical attributes, that is, general features of the object. For this view, see Chignell (2007) and Brown (2004).
8   For a similar reading, see also Kenneth Berry. He describes an aesthetic idea as a 'mental image' or a 'mind picture' (2008: 106). Similarly, Thomas Teufel explains an aesthetic idea as a 'totalizing vision' (2019: 3115).
9   Strictly speaking, empirical intuitions belong to the receptive faculty of sensibility; yet, Kant often uses the term in the broad sense to include synthesized (apprehended and reproduced) manifold of intuition, that is, images. See Matherne (2015).
10 For this view, see also Paton (1965), Pippin (1992), Ewing (1967), Guyer (2006) and Matherne (2015). The opposite view, namely, that formation of images can occur prior to conceptual synthesis, has been argued primarily by Young (1988), Gibbons (1994), Allison (2004) and Hanna (2005).
11 This is also pointed out by DeBort (2012).
12 One should bear in mind the distinction between abstractness and abstraction. Even though all concepts are abstracted from experience, not all of them are abstract concepts. Category members of abstract concepts are non-material, non-concrete and non-sensory-perceivable objects in contrast to superordinate concepts (like the concepts of animal, furniture, artefact and so forth), whose members are all concrete and sensory-perceivable objects. See Borghi and Binkofski (2014: 3–4).
13 Even though concepts of emotions such as 'love' or 'hopelessness' are usually included in the category of abstract concepts, since they both refer to non-concrete and non-physical objects, contemporary cognitive science and psycholinguistics tend to keep them apart. This is because emotion concepts (such as happiness, hopefulness, love, sadness, anxiety, jealousy and loneliness) refer directly to internal affective states and have a direct bodily counterpart, while abstract concepts lack

these two characteristics. Rather, they refer to mental states, cognitive processes, personality traits, situations and events (for example, concepts such as time, thought, death, truth, infinity, chaos, patriotism) which might have an indirect emotional or affective association (for example, the concept of death might be associated with negative emotional experiences such as fear). See Borghi and Binkofski (2014: 2–11). In this book I refer to both kinds of concept as abstract.

14 For a survey of these studies, see Wiemer-Hastings and Xu Xu (2005) and Kiefer and Harpaintner (2020).

15 For the view that the kind of knowledge that aesthetic ideas give rise to cannot be captured in terms of logical or determinate knowledge, see also Chaouli (2011), Makkreel (1990: 120–2) and Matherne (2013).

16 This idea has also been suggested, though not further developed by Rudolf Makkreel. As he writes, 'although such [aesthetic] ideas cannot enlarge concepts qua concepts, they broaden our interpretation of experience' (1990: 122).

17 As Anthony Savile writes, an aesthetic idea offers 'one possible way of thinking about [a rational idea], and may be said to be an expression or presentation of that idea or theme' (1987: 170).

18 In this respect my interpretation of Kant's theory of artworks as expressions of aesthetic ideas bears a similarity with R. G. Collingwood's account of artistic expression as an individualization, rather than mere description of emotion concepts. As he writes: 'The poet, therefore, in proportion as he understands his business, gets as far away as possible from merely labelling his emotions as instances of this or that general kind, and takes enormous pains to individualize them by expressing them in terms which reveal their difference from any other emotion of the same sort' (1938: 113).

19 For an interesting criticism of the view that aesthetic ideas function as metaphors, see also Jane Forsey. She interprets aesthetic ideas as 'hermeneutically richer, more ambiguous, than a simple juxtaposition of two clearly disjoint elements' (2004: 580).

Chapter 3

1 As most notably pointed out by Currie (1995).
2 I am referring to the notion of factual self-knowledge as used by Jopling (2000: 17).
3 While I agree with Lucy O'Brien (2017) and her view that narrative art generates so-called recognitional self-knowledge, rather than completely new self-knowledge, I disagree with her claim that such self-knowledge does not count as a substantial piece of self-knowledge. According to my view, narrative art gives us the opportunity not merely to confirm already-existent *explicit* beliefs, as O'Brien suggests, but rather to recognize existent *implicit* beliefs and other self-concepts

that comprise the background of our experiences. This is the meaning of self-knowledge as endorsed by psychotherapy.

4 Introspection differs from our ordinary awareness of mental states in that it is a deliberate and attentive awareness. That is, while we are conscious of most of our mental states by simply experiencing them, we become introspectively aware of them only when we intentionally observe and reflect on them. For more on the distinction between ordinary and introspective awareness, see Rosenthal (2000).

5 For example, Seymour Epstein explains Tolstoy's inner conflict accordingly: 'His achievements, although viewed as successes in his rational system, failed to fulfil a basic need or needs in his experiential system. His success, therefore, can be said to be success at the rational level but failure at the experiential level' (2003: 16). In other words, while Tolstoy's feeling of unhappiness is in disagreement with his rationally endorsed reasons, it is nonetheless in agreement with his unconscious reasons. His unhappiness is the result of an unfulfilled unconscious need (whatever that might be) and thus responsive to some sort of reasons. His feeling of unhappiness is thus an appropriate response even though it is not determined by rational deliberations. He should feel unhappy, because he has (unconscious) reasons for it. Just because we cannot explicitly point out what these reasons are, it does not mean they are not real and that our reaction is not an appropriate one.

6 A similar account of the first-person authority has been previously suggested by Charles Taylor. According to him, our mental states are shaped by self-articulations or self-interpretations that we consider as true, whereby the correctness of a given self-interpretations is determined through feelings. A given articulation is constitutive of our mental state insofar as we come to "'see-feel' that this is the right description," whereby we feel the accuracy of articulation insofar it 'makes them [mental states] clearer and more defined' (1985: 70–1). The decision as to whether the given interpretation fits with our own psychological states is up to us and our own feeling of appropriateness.

7 For the review of these studies, see Gertler (2011: 70–81).

8 For the review of a dual system theory, see Evans (2008).

9 An interesting study, performed by Wilson, Lindsey and Schooler (2000), revealed that people can have dual, independently existing attitudes towards one and the same object, one implicit (unconscious), the other explicit (conscious) attitude constructed on the basis of reasons. Both attitudes influence our existing responses and actions independently; yet, it is the implicit attitude that is more persistent and difficult to change.

10 For a more detailed analysis of various kinds of erroneous self-information obtained through the process of introspection, see Nisbett and Wilson (1977).

11 Besides, many research studies have revealed that third-person perspective can often lead to what is called self-fabrication, namely mistakenly inferring mental states from our behaviour that did not previously exist (Wilson 2002: 206). For

example, we can mistakenly conclude that we like another person just because we agreed to go out on a date with them, without taking into consideration other reasons that might impinge upon our decision to go out on a date, such as a desire to please our parents. Mistaken inferences often occur because of our inattentive self-reflection on the reasons for our responses and actions.

12 In philosophical aesthetics, the distinction between first- and third-person perspective on fictional characters is also known as the distinction between the internal and external perspective respectively. See Lamarque (1996), Curie (2010: 49–63) and Clifton (2016).
13 For a similar view see Amy Coplan (2004). She writes that narrative experience simultaneously involves first-person empathic perspective on characters, as well as third-person perspective, whereby we observe characters' experiences from the outside.
14 The main proponent of this position is Gaut (2007: 203–26). This is the position well argued for by many contemporary philosophers.
15 For this point see also Moran (1994), Robinson (2005: 108–17), Lazarus and Lazarus (1996: 129–36) and Frijda (1988: 352).
16 This idea has been pointed out by numerous research studies, which show that perception of emotional and psychological similarities (or dissimilarities) with fictional characters facilitates (or hinders) the activation of mental simulation. See Green (2004) and Hakemulder (2000: 70–3).
17 This is the value of narrative simulation as pointed out by Schwan (2013).
18 For the view that we experience ourselves in a non-narrative, rather than narrative way, see also Strawson (2004) and Vice (2003). Both authors criticize the idea that experiencing our own life as a narrative, that is, thinking of our own lives as having a story-like structure, promotes self-understanding. In contrast, they argue that seeing oneself in narrative terms has pernicious effects on one's own self for the reason that it leads to self-deception and confabulations as each narration of one's own past experiences and events necessarily creates alterations of these experiences. Artificially imposing certain form-structure into our life events can lead to dehumanization, falsification and inauthenticity.
19 See also Currie (2007), who emphasizes the role of causal explanations in artistic narratives.

Chapter 4

1 Both agreeable and beautiful art belong to the category of aesthetic art as they both have 'feeling of pleasure as its immediate aim.' (KU 5:305; 184) Yet, while the ground of pleasure in agreeable aesthetic art is 'representations as mere sensations' that only aim at enjoyment, the ground of pleasure in beautiful art

is representations as 'as kinds of cognition' (KU 5:305; 184). That is, in contrast to agreeable art whose aim is only to entertain, beautiful art has a cognitive function in that it 'promotes the cultivation of the mental powers for sociable communication' (KU 5:306; 185).

2. According to Kant, the ability to express aesthetic ideas is a special talent (or spirit) possessed by a genius. Thus, only intentionally spirit-filled artworks can actually express aesthetic ideas. Even though this idea appears to be inconsistent with Kant's remark that all beauty, thus also natural beauty, can express aesthetic ideas (KU 5:320; 197), this inconsistency can be resolved by taking into account Kant's claim that nature is beautiful if it looks like art (KU 5:306; 185). Thus, nature can be seen as expressive of aesthetic ideas only if viewed as intentionally produced. Considered by itself alone, nature cannot be seen as expressive of aesthetic ideas. For this view, see also Kemal (1997: 141) and Guyer (2021).

3. While Kant's claim that all artistic beauty presupposes a concept of a purpose and it is thus of adherent kind appears to be inconsistent with his earlier remark that non-representational art, such as music without words, is free beauties, I believe there is a way to reconcile Kant's confusing position regarding the status of such kind of art. Namely, it is true that strictly speaking non-representational art has adherent kind of beauty, since it is made with a certain purpose. However, since the purpose of such art is free or purposeless beauty itself, that is, to give satisfaction in virtue of its form alone, judging the beauty of such art is judging it freely. In other words, there is no difference between adherent and free aesthetic judgements in the case of non-representational art.

4. For example, Samantha Matherne (2014) claims that music can be appreciated as both agreeable and beautiful depending on the attitude or different ways we approach music. We can appreciate music based on the effect it has on our body; how it pleases our sense in which case we judge is as agreeable. However, we can appreciate the same piece of music based on its formal composition, in which case we appreciate it as beautiful. This view has been recently favoured by Young (2020) as well.

5. See also Butts (2000) and Kivy (1993) for this view. For a nice discussion of the resemblance theory of musical expressiveness, see Davies (1980).

6. For this view, see also Tuna (2019).

7. For this point, see Penny (2008).

8. This has been also pointed out by Guyer (2015: 233).

9. A similar argument against the possibility of absolute music representing specific emotions is made by Hanslick (1986: 9–11).

10. See Prinz (2004).

11. See also Jørgensen (2018: 29–31) and Dyck (2019) and their discussion of the distinction between voluntary and involuntary imagination in Kant's aesthetic theory.

Chapter 5

1. Here I am employing Nick Zangwill's (1995) distinction between substantive aesthetic properties (such as elegance, balance, unity, graceful, daintiness) and verdictive or evaluative aesthetic properties (such as beauty or aesthetic merit). Evaluative aesthetic properties are bearers of aesthetic value and signify work's aesthetic merit. To say that work is beautiful is to judge it as aesthetically meritorious. While substantive aesthetic properties are relevant to aesthetic value, they are not themselves evaluative. To describe a work as elegant does not necessarily mean that the work is evaluated as having an aesthetic merit, though elegance can contribute to the overall aesthetic evaluation.
2. For the review of different interpretations as to how exactly imagination can reflect on the object freely without being governed by any determinate concepts, see Guyer (2006).
3. According to Kant, mere tones, sounds, colours and textures are mere sensations (KU 5:224; 108) that we passively receive through sensibility. Since their apprehension does not require the activity of imagination and understanding, they are not the subject of pure aesthetic experience.
4. This proposal has also been suggested by Guyer (2002: 448).
5. A similar attempt, though with notable differences, of expanding Kant's conception of formalism has been made by Jenny McMahon (2010). She distinguishes between a thin and thick conception of Kant's aesthetic formalism, the former referring to the composition of perceptual features without conceptual engagement, whereas the latter, which she ultimately attributes to Kant, includes various concepts, particularly concepts of reason.
6. According to aesthetic theories of art, the value of an artwork lies in its aesthetic value, whereas an aesthetic value is a value that an object has in virtue of its capacity to elicit aesthetic pleasure of the beautiful. That is, an object has an aesthetic value in virtue of possessing pleasing aesthetic properties (beauty) and which themselves supervene on formal-perceptual properties. See Monroe Beardsley's (2019) aesthetic theory of art as traditionally defined or more contemporary definition as defended by Nick Zangwill (2007).
7. For a detailed discussion on how non-perceptual art (specifically conceptual art) challenges aesthetic theories of art, see Shelley (2003), Zangwill (2002), Lamarque (2007) and Schellekens (2007).
8. Spirited ugliness consists in the conflicting and incoherent combination of aesthetic attributes, resulting in a displeasing disharmony between the faculties of imagination and understanding, and in the incongruity of thoughts conveyed.
9. While similar suggestion has been made by Robert Yanal, his account significantly differs from mine in that he attributes beauty of an aesthetic idea to the expressive

content, rather than to formal properties. Thus, he distinguishes between formal and expressive beauty. He expresses disappointment towards Kant's doctrine of aesthetic ideas 'for it does not really integrate Kant's theory of fine art into his theory of pure or formal beauty' (1994: 178).

10 A fine example of such a nonsensical representation is Lewis Carroll's poem *Jabberwocky* (1871). The poem is a play of made-up words, which do not have a specific meaning and are artist's original inventions. Even though the words themselves may be said to be a product of artist's use of free imagination and therefore exhibit originality, their combination however carries no meaning.

11 For an interesting discussion of Kant's notion of original nonsense, see also Lewis (2005). He distinguishes two forms of originally nonsensical representation, namely one that exhibits genius without taste (high degree of imagination not constrained by understanding), the other that exhibits neither genius nor taste and is marked by thoughtless and pretentious attempt to imitate the originality of the artist.

12 For an interesting discussion on the recognitional role of the feeling of pleasure in aesthetic judgements of taste, see Rogerson (2009: 67).

Chapter 6

1 For the most recent interpretations that emphasize a cognitive role of aesthetic pleasure, see Hughes (2017), Merritt (2014), Heidemann (2016), Pillow (2006) and Gorodeisky (2018). For example, Fiona Hughes argues that the feeling of pleasure is a form of reflective awareness that functions as a 'reflection on the possibility of cognition without counting as an actual cognition' (2017: 383). Keren Gorodeisky maintains a similar view when she claims that pleasure functions as a form of self-reflection that involves 'a specific kind of rational causation' or 'rational structure' (2018: 176). A different strategy is taken by Kirk Pillow, who argues that Kant's conception of cognition must be considered more broadly, to include not merely determinate cognition, but also 'interpretative ordering of experience into meaningful wholes' (2006: 252). Interpretive cognition is conceptually indeterminate and takes place in both empirical concept acquisitions and aesthetic judgements.

2 For example, deep-sea animal called *fangtooth* is a perfect specimen of the natural kind to which it belongs, that is, it satisfies all the conditions required for an object to belong to this kind; yet, it is still ugly.

3 For this point, see also Pippin (1992). Furthermore, as pointed out by Paul Guyer (2006), the application of empirical concepts to the sensible manifold is necessary condition not merely for cognition of the object, but to have an experience of the object in the first place. In short, the argument is that a priori concepts

of understanding (i.e. categories), which are responsible for the possibility of experiencing objects in the first place (concept of a substance, causality, etc.), can be applied to the representation only through the assistance of empirical concepts (as categories are abstract concepts and thus cannot differentiate between various particular images and laws). This means that in order to have an experience of a particular object, the apprehension of the sensible manifold must be guided, not only by pure concepts, but by the particular empirical concepts as well. Empirical concepts are necessary for the experience of objects. Moreover, this shows that it is impossible to have a state of mind in which cognitive powers were in free harmony, that is, without the application of empirical concepts, as Kant seems to claim that takes place in judgements of the beautiful. In order to find an object beautiful and experience free harmony, we must in the first place have conceptual harmony which necessitates the experience of an object. Free harmony, as a determining basis of the aesthetic feeling of pleasure, has a cognitive component, not only because it involves the engagement of the faculties that are required for ordinary cognition, but also because it depends on empirical concept application to some extent. Thus, as proposed by Guyer's recent metacognitive interpretation, free harmony is a cognitive harmony exercised to a high degree, that is, which exhibits order or unity that extends beyond the unity necessary for the recognition of an object (Guyer 2005).

4   Similarly, Ted Kinnaman identifies Kant's notion of cognition in general with the 'subsumability in a philosophical system of nature' (2000: 294).

5   For the opposite view, namely that there are two different principles, one for empirical concept acquisition and another for judgements of taste, see Allison (2001: 62–4), Guyer (1997: 44–7), Rueger and Evren (2005), Caranti (2005: 364–74), Hughes (2007: 255–69) and Zuckert (2007: 65–82).

6   That it is the individual aspects of the object that are taken into consideration in aesthetic reflection is also pointed out by Cohen (2002), Zuckert (2006) and Gorodeisky (2011).

7   In fact, Kant claims that the feeling of pleasure also occurs in finding the empirical concept for the particular, yet, that such feeling of pleasure ceases to exist once it is 'mixed up with mere cognition and is no longer specially noticed' (KU 5:187; 74).

8   Harold Lee, for example, goes even further and identifies the feeling of pleasure with the 'universal concept which makes the reflective judgement possible' (1931: 541). Paul Gordon makes a similar suggestion when he characterizes beauty as a 'conceptless concept' (2015: 33).

9   Something similar is also suggested by Avner Baz, who writes that empirical judgements are 'guided by feeling – a sense of fit between concept and intuition', the difference being that in judgements of taste this feeling is made 'explicit' (2020: 81). Feeling of pleasure lies behind the possibility of cognition in general.

10 I believe something similar has Aaron Halper in mind, when he writes that judgements of taste are a 'subjective response to cognition' (2019: 42).
11 Elena Fell and Ioanna Kopsiafti characterize aesthetic cognition similarly, when they describe it as a 'unique form of sensuous or imaginatively intended knowledge, whose meanings can be experienced only through direct perceptual or imaginative engagement with the individual aesthetic or artistic object', whereby each aesthetic object communicates a unique meaning and a unique form of aesthetic cognition (2016: 1).

Bibliography

Abend, S. M. (2007), 'Therapeutic Action in Modern Conflict Theory', *Psychoanalytic Quarterly*, 76 (S): 1417–42.
Allison, H. (2001), *Kant's Theory of Taste: A Reading of the Critique of Aesthetic Judgment*, Cambridge: Cambridge University Press.
Allison, H. (2004), *Kant's Transcendental Idealism*, New Haven and London: Yale University Press.
Ames, D. R. (2004), 'Inside the Mind Reader's Tool Kit: Projection and Stereotyping in Mental State Inference', *Journal of Personality and Social Psychology*, 87 (3): 340–53.
Aquila, R. E. (1979), 'A New Look at Kant's Aesthetic Judgment', *Kant-Studien*, 70 (1): 17–43.
Aumann, A. (2014), 'The Relationship between Aesthetic Value and Cognitive Value', *The Journal of Aesthetics and Art Criticism*, 72 (2): 117–27.
Baccarini, E. and M. C. Urban (2013), 'The Moral and Cognitive Value of Art', *Ethics & Politics*, XV (1): 474–505.
Barsalou, L. W. and K. Wiemer-Hastings (2015), 'Situating Abstract Concepts', in D. Pecher and R. A. Zwaan (eds), *Grounding Cognition: The Role of Perception and Action in Memory, Language, and Thinking*, 129–63, Cambridge: Cambridge University Press.
Batson, D. C. (2009), 'Two Forms of Perspective Taking: Imagining How Another Feels and Imagining How You Would Feel', in K. D. Markman, W. M. P. Klein and J. A. Suhr (eds), *Handbook of Imagination and Mental Simulation*, 267–80, New York: Psychology Press.
Baumberger, C. (2011), 'Understanding and Its Relation to Knowledge', in C. Jäger and W. Löffler (eds), *Epistemology: Contexts, Values, Disagreement: Proceedings of the 34th International Ludwig Wittgenstein Symposium in Kirchberg*, 16–18, Kirchberg am Wechsel: Ontos Verlag.
Baumberger, C. (2013), 'Art and Understanding. In Defence of Aesthetic Cognitivism', in M. Greenlee, R. Hammwöhner, B. Köber, C. Wagner and C. Wolff (eds), *Bilder sehen. Perspektiven den Bildwissenschaft*, 41–67, Regensburg: Schnell und Steiner.
Baumberger, C. (2014), 'Types of Understanding: Their Nature and Their Relation to Knowledge', *Conceptus: Zeitschrift Fur Philosophie*, 40 (98): 67–88.
Baumberger, C., C. Beisbart and G. Brun (2017), 'What Is Understanding? An Overview of Recent Debates in Epistemology and Philosophy of Science', in S. Grimm, C. Baumberger and S. Ammon (eds), *Explaining Understanding. New Perspectives from Epistemology and Philosophy of Science*, 1–34, New York: Routledge.

Baz, A. (2005), 'Kant's Principle of Purposiveness and the Missing Point of (Aesthetic) Judgments', *Kantian Review*, 10 (1): 1–32.
Baz, A. (2020), *The Significance of Aspect Perception: Bringing the Phenomenal World into View*, Switzerland: Springer.
Beardsley, M. (2019), 'An Aesthetic Definition of Art', in P. Lamarque and S. H. Olsen (eds), *Aesthetics and the Philosophy of Art: The Analytic Tradition*, 22–9, New York: Wiley Blackwell.
Bell, D. and A. Leite (2016), 'Experiential Self-Understanding', *The International Journal of Psychoanalysis*, 97: 305–32.
Bem, D. J. (1972), 'Self-Perception Theory', in L. Berkowitz (ed), *Advances in Experimental Social Psychology*, 6, 1–62, New York: Academic.
Berger, D. (2009), *Kant's Aesthetic Theory: The Beautiful and Agreeable*, New York: Continuum.
Berry, K. (2008), 'Kandinsky, Kant, and Modern Mandala', *Journal of Aesthetic Education*, 42 (4): 105–10.
Bicknell, J. (2004), 'Self-Knowledge and the Limitations of Narrative', *Philosophy and Literature*, 28 (2): 406–16.
Bilgrami, A. (2006), *Self-Knowledge and Resentment*, Cambridge, MA: Harvard University Press.
Black, M. (1954–5), 'Metaphor', *Proceedings of the Aristotelian Society*, 55 (1): 273–94.
Borghi, A. and F. Binkofski (2014), *Words as Social Tools: An Embodied View on Abstract Concepts*, New York: Springer.
Brady, E. (2010), 'Ugliness and Nature', *Enrahonar: quaderns de filosofia*, 45: 27–40.
Brown, S. R. (2004), 'On the Mechanism of the Generation of Aesthetic Ideas in Kant's Critique of Judgment', *British Journal for the History of Philosophy*, 12 (3): 487–99.
Butts, R. E. (2000), 'Kant's Theory of Musical Sound: An Early Exercise in Cognitive Science', in S. Graham (ed), *Witches, Scientists, Philosophers: Essays and Lectures*, 107–26, Dordrecht: Kluwer Academic.
Cannon, J. (2008), 'The Intentionality of Judgments of Taste in Kant's Critique of Judgment', *The Journal of Aesthetics and Art Criticism*, 66 (1): 53–65.
Caranti, L. (2005), 'Logical Purposiveness and the Principle of Taste', *Kant-Studien*, 96 (3): 364–74.
Carman, T. (2003), 'First Persons: On Richard Moran's Authority and Estrangement', *An Interdisciplinary Journal of Philosophy*, 46 (3): 395–408.
Carroll, N. (1990), *The Philosophy of Horror: Or Paradoxes of the Heart*, London & New York: Routledge.
Carroll, N. (2002), 'The Wheel of Virtue: Art, Literature, and Moral Knowledge', *The Journal of Aesthetics and Art Criticism*, 60 (1): 3–26.
Carroll, N. (2018), 'Oedipus Tyrannus and the Cognitive Value of Literature', in P. Woodruff (ed), *The Oedipus Plays of Sophocles: Philosophical Perspectives*, 17–39, Oxford: Oxford University Press.

Carr, D. (1986), 'Narrative and the Real World: An Argument for Continuity', *History and Theory*, 25 (2): 117–31.

Chaouli, M. (2011), 'A Surfeit in Thinking: Kant's Aesthetic Ideas', *Yearbook of Comparative Literature*, 57: 55–77.

Chignell, A. (2007), 'Kant on the Normativity of Taste: The Role of Aesthetic Ideas', *Australian Journal of Philosophy*, 85 (3): 415–33.

Clifton, S. (2016), 'A Notorious Example of Failed Mindreading: Dramatic Irony and the Moral and Epistemic Value of Art', *The Journal of Aesthetic Education*, 50 (3): 73–90.

Coffa, A. J. (1991), *The Semantic Tradition from Kant to Carnap: To the Vienna Station*, Cambridge: Cambridge University Press.

Cohen, T. (2002), 'Three Problems in Kant's Aesthetics', *British Journal of Aesthetics*, 42 (1): 1–12.

Coleman, X. J. F. (1974), *The Harmony of Reason: A Study in Kant's Aesthetics*, Pittsburgh: University of Pittsburgh Press.

Collingwood, R. G. (1938), *The Principles of Art*, Oxford: Clarendon Press.

Coplan, A. (2004), 'Empathic Engagement with Narrative Fictions', *The Journal of Aesthetics and Art Criticism*, 62 (2): 141–52.

Coplan, A. (2006), 'Catching Characters' Emotions: Emotional Contagion Responses to Narrative Fiction Film', *Film Studies*, 8 (1): 26–38.

Coplan, A. (2011), 'Understanding Empathy: Its Features and Effects', in A. Coplan and P. Goldie (eds), *Empathy: Philosophical and Psychological Perspectives*, 3–18, Oxford: Oxford University Press.

Costello, D. (2013), 'Kant and the Problem of Strong Non-Perceptual Art', *The British Journal of Aesthetics*, 53 (3): 277–98.

Cooper, R. (2014), *Psychiatry and Philosophy of Science*, New York: Routledge.

Crawford, D.W. (1982), 'Kant's Theory of Creative Imagination', in T. Cohen and P. Guyer (eds), *Essays in Kant's Aesthetics*, 151–78, Chicago: University of Chicago Press.

Crawford, D. (2003), 'Kant's Theory of Creative Imagination', in P. Guyer (ed), *Kant's Critique of the Power of Judgment: Critical Essays*, 143–70, Lanham, MD: Rowman and Littlefield Publishers.

Currie, G. (1995), 'The Moral Psychology of Fiction', *Australian Journal of Philosophy*, 73 (2): 250–9.

Currie, G. (2007), 'Both Sides of the Story: Explaining Events in a Narrative', *Philosophical Studies: An International Journal for Philosophy in the Analytic Tradition*, 135 (1): 49–63.

Currie, G. (2010), *Narrators & Narrators: A Philosophy of Stories*, Oxford: Oxford University Press.

Currie, G. (2020), *Imagining and Knowing: The Shape of Fiction*, Oxford: Oxford University Press.

Currie, G. and I. Ravenscroft (2002), *Recreative Minds: Imagination in Philosophy and Psychology*, Oxford: Clarendon Press.

Dadlez, E. (2009), *Mirrors to Each Other*, Oxford: Wiley-Blackwell.
Davenport, E. A. (1983), 'Literature as Thought Experiment (on Aiding and Abetting the Muse)', *Philosophy of the Social Sciences*, 13 (93): 279–306.
Davies, D. (2010), 'Learning through Fictional Narratives in Art and Science', in R. Frigg and M. Hunter (eds), *Beyond Mimesis and Convention*, 51–71, Berlin: Springer.
Davies, S. (1980), 'The Expression of Emotion in Music', *Mind*, 89 (353): 67–86.
Davies, S. (1994), *Musical Meaning and Expression*, Ithaca and London: Cornell University Press.
Davidson, S. C. (2000), 'The Ambiguous Meaning of Musical Enchantment in Kant's Third Critique', in M. Kronegger and A. T. Tymieniecka (eds), *The Aesthetics of Enchantment in the Fine Arts*, 115–20, Boston and London: Kluwer Academic.
DeBort, C. (2012), 'Geist and Communication in Kant's Theory of Aesthetic Ideas', *Kantian Review*, 17 (2): 177–90.
Diffey, T. J. (1995), 'What Can We Learn from Art?', *Australian Journal of Philosophy*, 73 (2): 204–11.
Dorter, K. (1990), 'Conceptual Truth and Aesthetic Truth', *The Journal of Aesthetics and Art Criticism*, 48 (1): 37–51.
Dyck, C. W. (2019), 'Imagination and Association in Kant's Theory of Cognition', in R. Meer, G. Motta and G. Stiening (eds), *Konzepte der Einbildungskraft in der Philosophie, den Wissenschaften und den Künsten des 18. Jahrhunderts*, 351–70, Berlin and Boston: Walter de Gruyter.
Eagle, M. N. (2011), 'Classical Theory, the Enlightenment Vision, and Contemporary Psychoanalysis', in M. J. Diamond and C. Christian (eds), *The Second Century of Psychoanalysis: Evolving Perspectives on Therapeutic Action*, 41–67, London: Karnac Books.
Egan, D. (2016), 'Literature and Thought Experiments', *Journal of Aesthetics and Art Criticism*, 74 (2): 139–50.
Elgin, C. Z. (2002), 'Art in the Advancement of Understanding', *American Philosophical Quarterly*, 39 (1): 1–12.
Elgin, C. Z. (2007), 'The Laboratory of the Mind', in W. Huemer, J. Gibson and L. Pocci (eds), *A Sense of the World: Essays on Fiction, Narrative and Knowledge*, 43–54, London: Routledge.
Elgin, C. Z. (2009), 'Is Understanding Factive?', in A. Haddock, A. Millar and D. Pritchard (eds), *Epistemic Value*, 322–30, Oxford: Oxford University Press.
Elgin, C. Z. (2017), *True Enough*, Cambridge, MA: The MIT Press.
Elkin, J. (2004), *Pictures and Tears*, New York and London: Routledge.
Ellis, R. D. (1995), *Questioning Consciousness: The Interplay of Imagery, Cognition, and Emotion in the Human Brain*, Amsterdam: Benjamins.
Epley, N. and E. M. Caruso (2009), 'Perspective Taking: Misstepping into Others' Shoes', in K. D. Markman, W. M. P. Klein and J. A. Suhr (eds), *Handbook of Imagination and Mental Simulation*, 295–312, New York: Psychology Press.

Epstein, S. (2003), 'Cognitive-Experiential Self-Theory of Personality', in T. Millon and M. J. Lerner (eds), *Comprehensive Handbook of Psychology, Volume 5: Personality and Social Psychology*, 159–84, Hoboken, New Jersey: John Wiley & Sons.
Evans, J. S. B. T. (2008), 'Dual-Processing Accounts of Reasoning, Judgment, and Social Cognition', *Annual Review of Psychology*, 59: 255–78.
Ewing, A. C. (1967), *A Short Commentary on Kant's Critique of Pure Reason*, Chicago: University of Chicago Press.
Fell, E. and I. Kopsiafti (2016), *The Cognitive Basis of Aesthetics: Cassirer, Crowther, and the Future*, New York: Routledge.
Felski, R. (2008), *Uses of Literature*, Malden: Blackwell Publishing.
Feshbach, N. D. and S. Feshbach (2009), 'Empathy and Education', in J. Decety and W. Ickes (eds), *The Social Neuroscience of Empathy*, 85–98, Cambridge, MA: The MIT Press.
Finkelstein, D. H. (2003), *Expression and the Inner*, Cambridge: Harvard University Press.
Forsey, J. (2004), 'Metaphor and Symbol in the Interpretation of Art', *Contemporary Issues in Aesthetics*, 8 (3): 573–86.
Forrester, S. (2012), 'Why Kantian Symbols Cannot Be Kantian Metaphors', *Southwest Philosophy Review*, 28 (2): 107–27.
Fricke, C. (1990), 'Explaining the Inexplicable: The Hypotheses of the Faculty of Reflective Judgment in Kant's Third Critique', *Noûs*, 24 (1): 45–62.
Friend, S. (2007), 'Narrating the Truth (More) or Less', in M. Kieran and D. M. Lopez (eds), *Knowing Art: Essays in Aesthetics and Epistemology*, 35–50, Dordrecht: Springer.
Frijda, N. H. (1988), 'The Laws of Emotion', *American Psychologist*, 43 (5): 349–58.
Gabriel, G. (2013), 'Aesthetic Cognition and Aesthetic Judgment', in H. Hühn and J. Vigus (eds), *Symbol and Intuition: Comparative Studies in Kantian and Romantic-Period Aesthetics*, 185–90, New York: Modern Humanities Research Association and Routledge.
Gaskin, R. (2013), *Language, Truth, and Literature: A Defence of Literary Humanism*, Oxford: Oxford University Press.
Gaut, B. (2006), 'Art and Cognition', in M. Kieran (ed), *Contemporary Debates in Aesthetics and the Philosophy of Art*, 115–26, Malden: Blackwell Publishing.
Gaut, B. (2007), *Art, Emotion and Ethics*, Oxford: Oxford University Press.
Gendler, T. S. (2000), *Thought Experiment: On the Powers and Limits of Imaginary Cases*, New York: Garland Publishing.
Gendler, T. S. (2010), *Intuition, Imagination, and Philosophical Methodology*, Oxford: Oxford University Press.
Gendlin, E. T. (1964), 'A Theory of Personality Change', in P. Worchel and D. Byrne (eds), *Personality Change*, 100–48, New York: John Wiley & Sons.
Gendlin, E. T. (1965), 'The Discovery of Felt Meaning', in J. B. McDonald and R. R. Leeper (eds), *Language and Meaning*, Papers from the ASCD Conference, The Curriculum Research Institute, 45–62, Washington, DC: Association for Supervision and Curriculum Development.

Gendlin, E. T. (1968), 'The Experiential Response', in E. Hammer (ed), *Use of Interpretation in Treatment*, 208–27, New York: Grune & Stratton.

Gendlin, E. T. (1997), *Experiencing and the Creation of Meaning: A Philosophical and Psychological Approach to the Subjective*, Evanston, IL: Northwestern University Press.

Gertler, B. (2011), *Self-Knowledge*, New York: Routledge.

Gibbons, S. L. (1994), *Kant's Theory of Imagination: Bridging Gaps in Judgement and Experience*, Oxford: Clarendon Press.

Gibson, J. (2003), 'Between Truth and Triviality', *British Journal of Aesthetics*, 43 (3): 224–37.

Gibson, J. (2008), 'Cognitivism and the Arts', *Philosophy Compass*, 3 (4): 573–89.

Gibson, J. (2009), 'Literature and Knowledge', in R. Eldridge (ed), *Oxford Handbook of Philosophy and Literature*, 467–85, Oxford: Oxford University Press.

Ginsborg, H. (1990), 'Reflective Judgment and Taste', *Nous*, 24 (1): 63–78.

Ginsborg, H. (1997), 'Lawfulness without a Law: Kant on the Free Play of Imagination and Understanding', *Philosophical Topics*, 25 (1): 37–81.

Goldie, P. (2000), *The Emotions: A Philosophical Exploration*, Oxford: Oxford University Press.

Goodman, N. (1978), *Ways of Worldmaking*, Indianapolis: Hackett Publishing.

Gordon, P. (2015), *Art as the Absolute: Art's Relation to Metaphysics in Kant, Fichte, and Schopenhauer*, New York: Bloomsbury.

Gorodeisky, K. (2011), 'A Tale of Two Faculties', *British Journal of Aesthetics*, 51 (4): 415–36.

Gorodeisky, K. (2010), 'Schematizing without a Concept? Imagine That!' *Proceedings of the European Society for Aesthetics*, 2: 178–92.

Gorodeisky, K. (2018), 'Rationally Agential Pleasure? A Kantian Proposal', in L. Shapiro (ed), *Pleasure: A History*, 167–94, Oxford: Oxford University Press.

Gotshalk, D. W. (1967), 'Form and Expression in Kant's Aesthetics', *British Journal of Aesthetics*, 7 (3): 250–60.

Graham, G. (1995), 'Learning from Art', *British Journal of Aesthetics*, 35 (1): 26–37.

Graham, G. (2005), *Philosophy of the Arts*, London and New York: Routledge.

Green, M. C. (2004), 'Transportation into Narrative Worlds: The Role of Prior Knowledge and Perceived Realism', *Discourse Processes: A Multidisciplinary Journal*, 38 (2): 247–66.

Green, M. C. and J. K. Donahue (2009), 'Simulated Worlds: Transportation into Narratives', in K. D. Markman, W. M. P. Klein and J. A. Suhr (eds), *Handbook of Imagination and Mental Simulation*, 241–56, New York: Psychology Press.

Green, M. (2017), 'Narrative Fiction as a Source of Knowledge', in P. Olmos (ed), *Narration as Argument*, 47–67, Cham, Switzerland: Springer.

Grimm, S. R. (2012), 'The Value of Understanding', *Philosophy Compass*, 7 (2): 103–17.

Grimm, S. R. (2017), 'Understanding and Transparency', in S. R. Grimm, C. Baumberger and S. Ammon (eds), *Explaining Understanding: New Perspectives from Epistemology and Philosophy of Science*, 212–29, New York and London: Routledge.

Guyer, P. (1977), 'Formalism and Theory of Expressions in Kant's Aesthetics', *Kant-Studien*, 68 (1–4): 46–70.
Guyer, P. (1997), *Kant and the Claims of Taste*, Cambridge: Cambridge University Press.
Guyer, P. (2002), 'Beauty and Utility in Eighteenth-Century Aesthetics', *Eighteenth-Century Studies*, 35 (3): 439–53.
Guyer, P. (2005), 'Kant on the Purity of the Ugly', in P. Guyer (ed), *Values of Beauty: Historical Essays in Aesthetics*, 141–62, Cambridge: Cambridge University Press.
Guyer, P. (2006), 'The Harmony of the Faculties Revisited', in R. Kukla (ed), *Aesthetics and Cognition in Kant's Critical Philosophy*, 162–93, Cambridge: Cambridge University Press.
Guyer, P. (2015), 'Play and Society in the Lectures on Anthropology', in R. Clewis (ed), *Reading Kant's Lectures*, 223–42, Berlin and Boston: De Gruyter.
Guyer, P. (2021), 'Kant's Theory of Modern Art?' *Kantian Review*, 26 (4): 619–34.
Hakemulder, J. (2000), *The Moral Laboratory: Experiments Examining the Effects of Reading Literature on Social Perception and Moral Self-Concept*, Amsterdam and Philadelphia: John Benjamins Publishing Company.
Halper, A. (2019), 'Aesthetic Judgment as Parasitic on Cognition', *Kant Yearbook*, 11 (1): 41–59.
Halper, A. (2020), 'Rethinking Kant's Distinction between the Beauty of Art and the Beauty of Nature', *European Journal of Philosophy*, 28 (4): 857–75.
Hampshire, S. (1982), *Thought and Action*, Notre Dame: University of Notre Dame.
Hanna, R. (2005), 'Kant and Nonconceptual Content', *European Journal of Philosophy*, 13 (2): 247–90.
Hanslick, E. (1986), *On the Musically Beautiful: A Contribution towards the Revision of the Aesthetics of Music*, trans. G. Payzant, Indianapolis: Hackett Publishing Company.
Hatfield, E., J. T. Cacioppo and R. L. Rapson (1994), *Emotional Contagion (Studies in Emotion and Social Interaction)*, Cambridge: Cambridge University Press.
Heidemann, D. H. (2016), 'Kant's Aesthetic Nonconceptualism', in D. Schulting (ed), *Kantian Nonconceptualism*, 117–44, London: Palgrave Macmillan.
Helm, B. W. (2009), 'Emotions as Evaluative Feelings', *Emotion Review*, 1 (3): 248–55.
Horstmann, R. P. (1989), 'Why Must There Be a Transcendental Deduction in Kant's Critique of Judgment?' in E. Förster (ed), *Kant's Transcendental Deductions*, 157–76, Stanford: Stanford University Press.
Hughes, F. (2007), *Kant's Aesthetic Epistemology*, Edinburgh: Edinburgh University Press.
Hughes, F. (2017), 'Feeling the Life of the Mind: Mere Judging, Feeling, and Judgment', in M. C. Altman (ed), *The Palgrave Kant Handbook*, 381–406, London: Palgrave Macmillan.
Hunt, L. H. (2006), 'Motion Pictures as a Philosophical Resource', in N. Carroll and J. Choi (eds), *Philosophy of Film and Motion Pictures: An Anthology*, 397–405, Malden, Massachusetts: Blackwell Publishing.

Jackson, F. (1986), 'What Mary Didn't Know', *The Journal of Philosophy*, 83 (5): 291–5.
John, E. (2005), 'Literary Fiction and the Philosophical Value of Detail', in M. Kieran and D. M. Lopes (eds), *Imagination, Philosophy, and the Arts*, 142–59, New York and London: Routledge.
Johnson, M. (2007), *The Meaning of the Body: Aesthetics of Human Understanding*, Chicago: University of Chicago Press.
Johnstone, H. W. (1970), *The Problem of the Self*, University Park, PA: Pennsylvania State University Press.
Jones, S. (2017), 'Aesthetic Ineffability', *Philosophy Compass*, 12 (2): 1–12.
Jopling, D. A. (2000), *Self-Knowledge and the Self*, New York and London: Routledge.
Jørgensen, D. (2018), 'The Philosophy of Imagination', in T. Zittoun and V. P. Glaveanu (eds), *Handbook of Imagination and Culture*, 19–46, Oxford: Oxford University Press.
Kajtar, L. (2016), 'What Mary Didn't Read: On Literary Narratives and Knowledge', *Ratio: An International Journal of Analytic Philosophy*, 29 (3): 327–43.
Kemal, S. (1997), *Kant's Aesthetic Theory: An Introduction*, New York: St Martin's Press.
Kiefer, M. and M. Harpaintner (2020), 'Varieties of Abstract Concepts and Their Grounding in Perception or Action', *Open Psychology*, 2 (1): 119–37.
Kieran, M. (1996), 'Art, Imagination, and the Cultivation of Morals', *The Journal of Aesthetics and Art Criticism*, 54 (4): 337–51.
Kieran, M. (1997), 'Aesthetic Value: Beauty, Ugliness and Incoherence', *Philosophy*, 72 (281): 383–99.
Kinnaman, T. (2000), 'Symbolism and Cognition in General in Kant's Critique of Judgment', *Archiv für Geschichte der Philosophie*, 82 (3): 266–96.
Kivy, P. (1993), 'Kant and the Affektenlehre: What He Said, and What I Wish He Had Said', in P. Kivy (ed), *The Fine Art of Repetition: Essays in the Philosophy of Music*, 250–64, Cambridge: Cambridge University Press.
Kivy, P. (1997), 'The Laboratory of Fictional Truth', in P. Kivy (ed), *Philosophies of Arts: An Essay in Differences*, 120–39, Cambridge: Cambridge University Press.
Kloosterboer, N. (2015), 'Transparent Emotions? A Critical Analysis of Moran's Transparency Claim', *Philosophical Explorations*, 18 (2): 246–58.
Knight, D. (2006), 'In Fictional Shoes: Mental Simulation and Fiction', in N. Carroll and J. Choi (eds), *Philosophy of Film and Motion Pictures*, 271–80, Malden: Blackwell Publishing.
Konstan, D. (2015), 'Fiction's False Start', in A. C. Sukla (ed), *Fiction and Art: Explorations in Contemporary Theory*, 9–24, London and New York: Bloomsbury Academic.
Kousta, S. T., G. Vigliocco, D. P. Vinson, M. Andrews and E. D. Campo (2011), 'The Representation of Abstract Words: Why Emotion Matters', *Journal of Experimental Psychology*, 140 (1): 14–34.
Kuiken, D., L. Phillips, M. Gregus, D. S. Miall, M. Verbitsky and A. Tonkonogy (2004), 'Locating Self-Modifying Feelings within Literary Reading', *Discourse Processes*, 38 (2): 267–86.

Küplen, M. (2015), *Beauty, Ugliness and the Free Play of Imagination: An Approach to Kant's Aesthetics*, New York: Springer.
Kvanvig, J. L. (2003), *The Value of Knowledge and the Pursuit of Understanding*, Cambridge: Cambridge University Press.
Lamarque, P. (1996), *Fictional Points of View*, New York: Cornell University Press.
Lamarque, P. (2004), 'On Not Expecting Too Much from Narrative', *Mind & Language*, 19 (4): 393–408.
Lamarque, P. (2006), 'Cognitive Values in the Arts: Marking the Boundaries', M. Kieran (ed), *Contemporary Debates in Aesthetics and the Philosophy of Art*, 127–39, Malden: Blackwell Publishing.
Lamarque, P. (2007), 'On Perceiving Conceptual Art', in P. Goldie and E. Schellekens (eds), *Philosophy and Conceptual Art*, 3–17, Oxford: Oxford University Press.
Lamarque, P. (2009), *The Philosophy of Literature*, Oxford: Blackwell.
Lamarque, P. (2010), 'Precis of the Philosophy of Literature', *British Journal of Aesthetics*, 50 (1): 77–80.
Lamarque, P. (2014), *The Opacity of Narrative*, London: Rowman and Littlefield.
Lamarque, P. and S. H. Olsen (1994), *Truth, Fiction, and Literature: A Philosophical Perspective*, Oxford: Clarendon Press.
Lazarus, R. S. and B. N. Lazarus (1996), *Passion and Reason: Making Sense of Our Emotions*, Oxford: Oxford University Press.
Lear, J. (2004), 'Avowal and Unfreedom', *Philosophy and Phenomenological Research*, LXIX (2): 448–54.
Lee, H. N. (1931), 'Kant's Theory of Aesthetics', *The Philosophical Review*, 40 (6): 537–48.
Lewis, P. (2005), 'Original Nonsense: Art and Genius in Kant's Aesthetics', in G. M. Ross and T. McWalter (eds), *Kant and His Influence*, 126–45, New York: Continuum International Publishing Group.
Lorand, R. (1994), 'Beauty and Its Opposites', *The Journal of Aesthetics and Art Criticism*, 52 (4): 399–406.
Lyotard, J. F. (1994), *Lessons on the Analytic of the Sublime*, trans. E. Rottenberg, Stanford: Stanford University Press.
Lüthe, R. (1984), 'Kants Lehre von den ästhetischen Ideen', *Kant-Studien*, 75: 65–74.
Makkreel, R. A. (1990), *Imagination and Interpretation in Kant: The Hermeneutical Import of the Critique of Judgment*, Chicago: University of Chicago Press.
Makkreel, R. A. (2006), 'Reflection, Reflective Judgment and Aesthetic Exemplarity', in R. Kukla (ed), *Aesthetics and Cognition in Kant's Critical Philosophy*, 223–44, Cambridge: Cambridge University Press.
Mar, R. A. and K. Oatley (2008), 'The Function of Fiction Is the Abstraction and Simulation of Social Experience', *Perspectives on Psychological Science*, 3 (3): 173–92.
Matherne, S. (2013), 'The Inclusive Interpretation of Kant's Aesthetic Ideas', *British Journal of Aesthetics*, 53 (1): 21–39.

Matherne, S. (2014), 'Kant's Expressive Theory of Music', *The Journal of Aesthetics and Art Criticism*, 72 (1): 129–45.
Matherne, S. (2015), 'Images and Kant's Theory of Perception', *Ergo*, 2 (29): 737–77.
Matheson, C. and E. Kirchhoff (2003), 'Critical Notice of James O. Young, Art and Knowledge', *Canadian Journal of Philosophy*, 33 (4): 575–98.
Matthews, P. M. (2010), *The Significance of Beauty: Kant on Feeling and the System of the Mind*, Dordrecht: Kluwer Academic Publishers.
Marres, R. (1989), *In Defense of Mentalism: A Critical Review of the Philosophy of Mind*, Amsterdam: Rodopi.
Merritt, M. M. (2014), 'Kant on the Pleasures of Understanding', in A. Cohen (ed), *Kant on Emotion and Value*, 126–45, New York: Palgrave Macmillan.
Mew, P. (1973), 'Facts in Fiction', *The Journal of Aesthetics and Art Criticism*, 31 (3): 329–37.
McDowell, J. (1996), *Mind and World*, Harvard: Harvard University Press.
McGeer, V. (2008), 'The Moral Development of First-Person Authority', *European Journal of Philosophy*, 16 (1): 81–108.
McMahon, J. (2010), 'The Classical Trinity and Kant's Aesthetic Formalism', *Critical Horizons: A Journal of Philosophy and Social Theory*, 11 (3): 419–41.
McMahon, J. A. (2017), 'Immediate Judgment and Non-Cognitive Ideas: The Pervasive and Persistent in the Misreading of Kant's Aesthetic Formalism', in M. C. Altman (ed), *The Palgrave Kant Handbook*, 425–46, London: Palgrave Macmillan.
Mikkonen, J. (2013), *The Cognitive Value of Philosophical Fiction*, London and New York: Bloomsbury.
Mikkonen, J. (2015), 'On Studying the Cognitive Value of Literature', *Journal of Aesthetics and Art Criticism*, 73 (3): 273–82.
Mikkonen, J. (2021), *Philosophy, Literature and Understanding: On Reading and Cognition*, London and New York: Bloomsbury.
Moore, A. W. (2007), 'Is the Feeling of Unity That Kant Identifies in His Third Critique a Type of Inexpressible Knowledge?', *Philosophy*, 82 (3): 475–85.
Moran, R. (1994), 'The Expression of Feeling in Imagination', *The Philosophical Review*, 103 (1): 75–106.
Moran, R. (2001), *Authority and Estrangement: An Essay on Self-Knowledge*, Princeton: Princeton University Press.
Neill, A. (2006), 'Empathy and (Film) Fiction', in N. Carroll and J. Choi (eds), *Philosophy of Film and Motion Pictures*, 247–59, Malden: Blackwell Publishing.
Neville, M. R. (1974), 'Kant's Characterization of Aesthetic Experience', *The Journal of Aesthetics and Art Criticism*, 33 (2): 193–202.
New, C. (1999), *Philosophy of Literature: An Introduction*, London: Routledge.
Nickerson, R. S. (1999), 'How We Know – and Sometimes Misjudge – What Others Know: Imputing One's Own Knowledge to Others', *Psychological Bulletin*, 125 (6): 737–59.

Nickerson, R. S., S. F. Butler and M. Carlin (2009), 'Empathy and Knowledge Projection', in J. Decety and W. Ickes (eds), *The Social Neuroscience of Empathy*, 43–56, Cambridge, MA: The MIT Press.

Nisbett, R. E. and T. D. Wilson (1977), 'Telling More than We Can Know: Verbal Reports on Mental Processes', *Psychological Review*, 84: 231–59.

Novitz, D. (2004), 'Knowledge and Art', in I. Niiniluoto, M. Sintonen and K. Wolenski (eds), *Handbook of Epistemology*, 985–1012, Dordrecht: Springer.

Nussbaum, M. C. (1990), *Love's Knowledge: Essays on Philosophy and Literature*, Oxford: Oxford University Press.

Nussbaum, M. C. (2001), *Upheavals of Thought: The Intelligence of Emotions*, Cambridge: Cambridge University Press, 183–99.

Nussbaum, M. (2004), 'Emotions as Judgments of Value and Importance', in R. C. Solomon (ed), *Thinking about Feeling: Contemporary Philosophers on Emotions*, 183–99, Oxford: Oxford University Press.

Nuyen, A. T. (1989), 'The Kantian Theory of Metaphor', *Philosophy & Rhetoric*, 22 (2): 95–109.

Nuzzo, A. (2005), *Kant and the Unity of Reason*, West Lafayette, IN: Purdue University Press.

O'Brien, L. (2017), 'The Novel as a Source of Self-Knowledge', in E. S. Bisset, H. Bradley and P. Noordhof (eds), *Art and Belief*, 135–50, Oxford: Oxford University Press.

Ortony, A. (1975), 'Why Metaphors Are Necessary and Not Just Nice', *Educational Theory*, 25 (1): 45–53.

Paivio, A. (1990), *Mental Representations: A Dual Coding Approach*, Oxford: Oxford University Press.

Palmer, L. (2011), 'On the Necessity of Beauty', *Kant-Studien*, 102 (3): 350–66.

Parret, H. (1998), 'Kant on Music and the Hierarchy of the Arts', *The Journal of Aesthetics and Art Criticism*, 56 (1): 251–64.

Parrott, M. (2015), 'Expressing First-Person Authority', *Philosophical Studies*, 172 (8): 2215–37.

Paton, H. J. (1965), *Kant's Metaphysic of Experience*, New York: The Humanities Press.

Peels, R. (2022), 'How Literature Delivers Knowledge and Understanding, Illustrated by Hardy's Tess of the D'Urbervilles and Wharton's Summer', *British Journal of Aesthetics*, 60 (2): 199–222.

Penny, L. (2008), 'The Highest of All the Arts: Kant and Poetry', *Philosophy and Literature*, 32 (1): 373–84.

Phelan, J. W. (2021), *Literature and Understanding: The Value of a Close Reading of Literary Texts*, London and New York: Routledge.

Pillow, K. (2001), 'Jupiter's Eagle and the Despot's Hand Mill: Two Views on Metaphor in Kant', *The Journal of Aesthetics and Art Criticism*, 59 (1): 193–209.

Pillow, K. (2006), 'Understanding Aestheticized', in R. Kukla (ed), *Aesthetics and Cognition in Kant's Critical Philosophy*, 245–65, Cambridge: Cambridge University Press.

Pippin, R. (1992), 'The Schematism and Empirical Concepts', in R. Chadwick and C. Cazeaux (eds), *Immanuel Kant: Critical Assessments*, 286–303, London and New York: Routledge.

Prinz, J. (2004), 'Embodied Emotions', in R. C. Solomon (ed), *Thinking about Feeling: Contemporary Philosophers on Emotions*, 44–60, Oxford: Oxford University Press.

Raffman, D. (1993), *Language, Music, and Mind*, Cambridge, MA: The MIT Press.

Reimer, B. (1986), 'The Nonconceptual Nature of Aesthetic Cognition', *The Journal of Aesthetic Education*, 20 (4): 111–17.

Reiter, A. (2017), 'Kant and Hegel on the End and Means of Fine Art: From an A-Historical to a Historical Conception of Art', *Hegel-Jahrbuch*, 1: 35–40.

Richards, I. A. (2004), *Principles of Literary Criticism*, London and New York: Routledge.

Robinson, J. (1995), 'L'éducation sentimentale', *Australasian Journal of Philosophy*, 73 (2): 212–26.

Robinson, J. (2005), *Deeper than Reason: Emotion and Its Role in Literature, Music, and Art*, Oxford: Oxford University Press.

Rogerson, K. F. (1986), *Kant's Aesthetics: The Roles of Form and Expression*, Lanham, MD: University Press of America.

Rogerson, K. F. (1998), 'Pleasure and Fit in Kant's Aesthetics', *Kantian Review*, 2: 117–33.

Rogerson, K. F. (2009), *The Problem of Free Harmony in Kant's Aesthetics*, New York: State university of New York Press.

Rosenthal, D. (2000), 'Introspection and Self-Interpretation', *Philosophical Topics*, 28 (2): 201–33.

Rueger, A. (2008), 'The Free Play of the Faculties and the Status of Natural Beauty in Kant's Theory of Taste', *Archiv für Geschichte der Philosophie*, 90 (3): 298–322.

Rueger, A. and S. Evren (2005), 'The Role of Symbolic Presentation in Kant's Theory of Taste', *British Journal of Aesthetics*, 45 (3): 229–47.

Russell, B. (2006), 'The Philosophical Limits of Film', in N. Carroll and J. Choi (eds), *Philosophy of Film and Motion Pictures: An Anthology*, 387–90, Malden, MA: Blackwell Publishing.

Sassen, B. (2003), 'Artistic Genius and the Question of Creativity', in P. Guyer (ed), *Kant's Critique of the Power of Judgment: Critical Essays*, 171–80, Lanham, MD: Rowman and Littlefield Publishers.

Savile, A. (1987), *Aesthetic Reconstructions: The Seminal Writings of Lessing, Kant, and Schiller*, Oxford: Blackwell.

Savile, A. (1993), *Kantian aesthetics pursued*, Edinburgh: Edinburgh University Press.

Schellekens, E. (2007), 'The Aesthetic Value of Ideas', in P. Goldie and E. Schellekens (eds), *Philosophy and Conceptual Art*, 71–91, Oxford: Oxford University Press.

Schick, T. W. (1982), 'Can Fictional Literature Communicate Knowledge?' *Journal of Aesthetic Education*, 16 (2): 31–9.

Schueller, H. M. (1955), 'Immanuel Kant and the Aesthetics of Music', *The Journal of Aesthetics and Art Criticism*, 14 (1): 218–47.

Schwanenflugel, P. J. (1991), *The Psychology of Word Meanings*, Hillsdale, NJ: Erlbaum.

Schwan, S. (2013), 'The Art of Simplifying Events', in A. P. Shimamura (ed), *Psychocinematics: Exploring Cognition as the Movies*, 214–26, Oxford: Oxford University Press.

Shelley, J. (2003), 'The Problem of Non-Perceptual Art', *British Journal of Aesthetics*, 43 (4): 363–78.

Smith, M. (2006), 'Film Art, Argument, and Ambiguity' *The Journal of Aesthetics and Art Criticism*, 64 (1): 33–42.

Solomon, R. C. (2007), *True to Our Feelings: What Our Emotions Are Really Telling Us*, Oxford: Oxford University Press.

Sorensen, R. (1998), *Thought Experiments*, Oxford: Oxford University Press.

Spackman, J. (2012), 'Expressiveness, Ineffability, and Nonconceptuality', *The Journal of Aesthetics and Art Criticism*, 70 (3): 303–14.

Stolnitz, J. (1992), 'On the Cognitive Triviality of Art', *The British Journal of Aesthetics*, 32 (3): 191–200.

Strawson, G. (2004), 'Against Narrativity', *Ratio*, XVII (4): 428–52.

Strijbos, D. and F. Jongepier (2018), 'Self-Knowledge in Psychotherapy: Adopting a Dual Perspective on One's Own Mental States', *Philosophy, Psychiatry and Psychology*, 25 (1): 45–58.

Stroud, S. T. (2008), 'Simulation, Subjective Knowledge, and the Cognitive Value of Literary Narrative', *Journal of Aesthetic Education*, 42 (3): 19–41.

Swirski, P. (2007), *Of Literature and Knowledge: Explorations in Narrative Thought Experiments, Evolution, and Game Theory*, London and New York: Routledge.

Taylor, C. (1985), *Human Agency and Language: Philosophical Papers*, Cambridge: Cambridge University Press.

Teufel, T. (2019), '"Much That Is Unnameable" in Nature and in Art: Kant's Doctrine of Aesthetic Ideas', in V. Waibel and M. Ruffing (eds), *Natur und Freiheit: Akten des XII. Internationalen Kant-Kongresses*, 3113–21, Berlin: De Gruyter.

Tinguely, J. J. (2013), 'Kantian Meta-Aesthetics and the Neglected Alternative', *British Journal of Aesthetics*, 53 (2): 211–35.

Tolstoy, L. (1983), *Confession*, New York and London: W. W. Norton & Company, Inc.

Tuna, E. H. (2016), 'A Kantian Hybrid Theory of Art Criticism: A Particularist Appeal to the Generalists', *Journal of Aesthetics and Art Criticism*, 74 (1): 397–411.

Tuna, E. H. (2019), 'Why Didn't Kant Think Highly of Music?', in V. Waibel and M. Ruffing (eds), *Natur und Freiheit: Akten des XII. Internationalen Kant-Kongresses*, 3141–8, Berlin: De Gruyter.

Uehling, T. E. (1971), *The Notion of Form in Kant's Critique of Aesthetic Judgment*, The Hague: Mouton.

Varga, S. (2015), *Naturalism, Interpretation, and Mental Disorder*, Oxford: Oxford University Press.

Vice, S. (2003), 'Literature and the Narrative Self', *Philosophy*, 1 (1): 93–108.

Walsh, D. (1969), *Literature and Knowledge*, Middletown: Wesleyan University Press.

Wartenberg, T. E. (2007), *Thinking on Screen: Film as Philosophy*, London and New York: Routledge.
Weatherston, M. (1996), 'Kant's Assessment of Music in The Critique of Judgment', *British Journal of Aesthetics*, 36 (1): 56–65.
Weitz, M. (1943), 'Does Art Tell the Truth?', *Philosophy and Phenomenological Research*, 3 (3): 338–48.
Wicks, R. (2015), 'The Divine Inspiration for Kant's Formalist Theory of Beauty', *Kant Studies Online*, 1: 1–31.
Wiemer-Hastings, K. and X. Xu (2005), 'Content Differences for Abstract and Concrete Concepts', *Cognitive Science*, 29 (5): 719–36.
Wilson, C. (1983), 'Literature and Knowledge', *Philosophy*, 58 (226): 489–96.
Wilson, T. D. (2002), *Strangers to Ourselves: Discovering the Adaptive Unconscious*, Cambridge, MA: The Belknap Press of Harvard University Press.
Wilson, T. D. and E. W. Dunn (2004), 'Self-Knowledge: Its Limits, Value, and Potential for Improvement', *Annual Review of Psychology*, 55: 493–518.
Wilson, T. D., S. Lindsey and T. Y. Schooler (2000), 'A Model of Dual Attitudes', *Psychological Review*, 107 (1): 101–26.
Wilkes, K. V. (1988), *Real People: Personal Identity without Thought Experiments*, Oxford: Oxford University Press.
Yanal, R. J. (1994), 'Kant on Aesthetic Ideas and Beauty', in R. J. Yanal (ed), *Institutions of Art: Reconsiderations of George Dickie's Philosophy*, 157–84, University Park, PA: The Pennsylvania State University Press.
Young, J. M. (1988), 'Construction, Schematism, and Imagination', *Topoi*, 3 (2): 123–31.
Young, J. O. (2001), *Art and Knowledge*, London and New York: Routledge.
Young, J. O. (2020), 'Kant's Musical Antiformalism', *The Journal of Aesthetics and Art Criticism*, 78 (2): 171–81.
Zagzebski, L. (2001), 'Recovering Understanding', in M. Steup (ed), *Knowledge, Truth, and Duty: Essays on Epistemic Justification, Responsibility, and Virtue*, 235–52, Oxford: Oxford University Press.
Zangwill, N. (1995), 'The Beautiful, the Dainty and the Dumpy', *British Journal of Aesthetics*, 35 (4): 317–29.
Zangwill, N. (2002), 'Are There Counterexamples to Aesthetic Theories of Art', *The Journal of Aesthetics and Art Criticism*, 60 (2): 111–18.
Zangwill, N. (2007), *Aesthetic Creation*, Oxford: Oxford University Press.
Zuckert, R. (2006), 'The Purposiveness of Form: A Reading of Kant's Aesthetic Formalism', *Journal of the History of Philosophy*, 44 (4): 599–622.
Zuckert, R. (2007), *Kant on Beauty and Biology: An Interpretation of the Critique of Judgment*, Cambridge: Cambridge University Press.

Index

abstract art 8, 98–9, 117–18
abstract concept 7–8, 49–50, 59–62, 64–9, 75–6, 124, 136, 144, 153, 162
abstract phenomena 59, 61–2, 102, 114, 140–1, 166–9
aesthetic anti-cognitivism 1, 2, 5–6, 11, 165, 169
aesthetic appreciation 4–5, 122–3, 126, 133, 168–9
aesthetic attribute 49–59, 68–9, 102–4, 131–3, 138–41, 143–4, 153–5, 166–8
See aesthetic ideas
aesthetic cognitivism 1–2, 5–9, 11–13, 18–19, 37, 39, 45, 121–5, 165–8, 194
aesthetic enjoyment 5, 18
aesthetic experience 6, 59, 116, 130, 142
aesthetic form 6, 125, 131, 139, 161
aesthetic ideas 7–8, 47–9, 53–4, 57–9, 64–73, 75–6, 97–106, 108–10, 113–19, 121–5, 129–32, 134–5, 137–8, 140–1, 166–7
aesthetic reflection 116, 152, 155, 157, 160–2
aesthetic value 4–5, 47–8, 59, 64–70, 102–3, 121–2, 143, 151–2, 159–62, 168–9, 171–6, 179–88
aesthetic-relevance objection 12, 18
affect 26, 85, 100, 104, 106, 115, 159
agential authority 78–9, 84
agreeable 53, 98–9, 105, 113–14, 117
alienation 7, 47, 49, 68, 117–18, 167
apprehension 7, 47, 55, 59, 127, 129, 142
association 8, 50, 54–5, 59–60, 71–2, 103–9, 114–16, 119, 138
associational thoughts 52, 54, 56, 71–2, 102, 104, 111, 116, 125, 132, 139, 141, 144, 155, 167, 168
attunement 131, 140

Baumberger, Christoph 40, 43–4
beauty
    adherent beauty 99–100, 128
    free beauty 99, 124, 158
    perceptual beauty 125, 132–5, 137–8
    spirited beauty 122, 125, 131–8
belief 5, 12–13, 33, 41–2, 79, 80, 83, 86, 92, 113

Cage, John 134
Cannon, Joseph 158
Carroll, Noel 16, 26, 32
Chignell, Andrew 54, 131
Coffa, J. Alberto 110
cognitive experience 5–6, 18, 42
cognitive judgment 142–3, 145, 156
cognitive triviality 7, 11, 23, 32
cognitive value 2–9, 11–14, 17–19, 21–4, 32, 35, 43–5, 59, 97–8, 100–3, 114, 118, 121–5, 140–1, 165–9, 171–3, 175–6, 181–3, 186–7, 194
color 28, 49, 100–2, 104, 108, 116–17, 136, 151, 153–4, 163
Coplan, Amy 25, 30, 91

determinate cognition 50, 127, 144–7, 149, 161–2
determinate concept 7, 48, 50, 54–5, 59–60, 142–4, 152–3, 155, 168

Egan, David 33
Elgin, Catherine 40, 41, 43, 44, 166
Elkins, James 116
emotion 27–8, 61–5, 78, 81, 87, 91–3, 106, 110, 113
emotional asymmetry 26
emotional contagion 27–9
emotional education 5, 18
emotional engagement 19, 26, 28, 34, 77
emotional experience 76, 82, 88–91
emotional state 49, 67, 86, 94, 154, 167
empathic inaccuracy 30–1 *See* empathy
empathy 21, 22, 26–8, 30, 75
empirical concept 49–51, 59–60, 150–1, 184

empirical intuition 48, 54, 55–7, 107–9, 111, 132, 135–6, 138, 140
epistemic commitment 44–5 *See* Elgin, Catherine
epistemic thesis 7, 121, 123
exemplification 44–5
experience-related properties
    introspective, emotional, affective 7, 61–2, 163, 167
experiential information 47, 49, 62, 65, 67, 69, 167
expressive 8, 18, 79, 97–8, 100–1, 112–13, 118–19, 124–5, 129–30, 151

factual truth 12, 19, 39–40, 69, 165, 169
fantasies 14–15, 114, 116–17
feeling of pleasure 6, 9, 43, 59, 72–3, 122, 125–6, 128, 132–3, 138–45, 152–60, 166–9
fiction 24, 32, 87
first-person authority 76, 78–81, 84, 92, 94–5
first-personal perspective 78, 80, 82–3, 85, 89, 91, 94
free harmony 8–9, 58, 72, 125–7, 129–32, 140–5, 147, 152–3, 156–61, 169
free play of imagination 51–2, 57, 106, 117, 126–7, 129–30, 137–9, 147, 177, 185
Friend, Stacie 13, 19

Gendlin, Eugene 61, 63–4
genius 58, 110, 137–9
Gibson, John 7, 17, 19, 23–4, 35, 84, 166
    See axiological understanding
Goldie, Peter 28
Guyer, Paul 125, 130, 138, 174, 178, 185

Haneke, Michael 67–68

imagination
    free imagination 100, 103, 110, 129, 155
    productive imagination 70, 102, 105–6, 108–11, 114–16, 139
    reproductive 48, 50–9, 64, 70, 72, 94, 97–103, 105–12, 114–18, 126–9, 135–6, 138–47, 152–3, 155–71
introspection 7, 62–3, 77–9, 83, 106
    See self-knowledge

Johnson, Mark 62
judgment
    aesthetic judgment 126–7, 130, 137, 156
    cognitive 127, 142–5, 147–52, 155–60, 162, 167
    judgment of taste 143–5, 147, 149–52, 155–60

knowledge
    conceptual knowledge 33, 35, 43, 165
    experiential knowledge 20–5, 29, 31, 165–6, 169
    philosophical knowledge 11, 32, 35
    propositional knowledge 1–6, 11–13, 15, 17, 19–27, 35–44, 75–7, 83–4, 121, 137, 151, 160–1, 165–6, 171, 177, 180, 183, 187, 194

Lamarque, Peter 2, 12, 39, 89
Landau, Sigalit 153, 162
literary art 25, 29, 35, 53, 97, 109–11, 115, 132, 135
live hypotheses 19
logical attributes 50, 52, 54, 58, 64, 140

Makkreel, Rudolf 160, 162
manifold of intuition 55–6, 126–7, 143–4, 146, 148–9
Matherne, Samantha 66
McGeer, Victoria 83
McMahon, Jennifer 54
memory 21, 66, 82, 107–8
mental life 8, 76–7, 79, 82, 84, 86, 88
mental representation 51–4, 59, 103, 105, 111, 114, 117, 131, 141, 163, 167
mental simulation 20–5, 84, 91–2, 94, 171
    See experiential knowledge
metaphor 71–2
Mikkonen, Jukka 43–4
mood 52, 64, 82, 162–3
Moran, Richard 79, 173
Munch, Edvard 28, 68, 118
music 8, 24, 40, 98–103, 105, 109, 112–15, 118, 124, 128

narrative art 87–9, 91
narrative content 34, 35, 43
no-justification objection 11, 15, 23, 32, 168

non-fictional artwork 13, 15–16
nonsense 140
Nussbaum, Martha 90

Olsen, Stein 2, 12
originality 3, 5, 67, 134, 137

paradox of fiction 13–14, 17, 22, 32, 166
perception 23, 29–30, 44, 83, 94, 112, 116, 126, 127, 151–2, 156, 161–2, 172, 179–80
perceptual form 124, 130, 137
personal life concerns 87–8, 113
pictorial art 28, 111–12, 115, 135
Pillow, Kirk 69–72, 102
plastic art 98, 115, 139
poetic art 109–10
poetry 53, 101, 109
principle of purposiveness 149–50, 152
productive aesthetic idea 8, 70, 97, 102–3, 105–16, 139, 158
psychological situation 85–6, 87, 93
psychotherapy 78, 81

rational ideas 48–50, 54, 57, 59, 64, 66, 69, 75, 100, 121–2, 124, 136, 141, 144, 162
real-life narrative 89–90
reflection 70, 75, 87, 90, 101–5, 110, 115, 119, 129, 145, 149, 150–3, 160–2
reflective judgment 72, 139, 148–9, 176
Reimer, Bennet 160
Robinson, Jenefer 17, 87
Rothko, Mark 108, 116

self-concept 75, 179
self-development 8, 69, 84, 95, 97
self-interpretation 78, 81
self-knowledge
    first-personal 20, 76–80, 82–3, 85–6, 89, 91, 94
    therapeutic 8, 21, 69, 75–7, 82, 84, 89, 90, 93–5
    third-personal 76–7, 86, 88–9, 91
self-oriented perspective 20–1, 28–30
semantic 6–7, 71–2, 110–11, 118–19, 130–1, 153, 161, 166–7, 169
sensation 8, 52–3, 64, 97–101, 103–6, 112–17, 128, 135, 140, 147, 159, 162

sense impressions 55–6, 109, 138, 143
sensible manifold 50–2, 56–7, 126–9, 142–3, 148–50, 152–3, 157–9
shape 38, 49, 51, 64–5, 68, 76, 78–9, 81, 86, 108, 112, 115, 118, 128, 136, 157, 158
sound 24, 104, 108, 132, 151
spatial 99, 112, 115, 128, 129
spirit 37, 53, 57, 99–101, 103, 110, 122–5, 134–5, 139
spiritless 123
Stolnitz, Jerome 1, 15
subject matter 3–5, 40, 42–3, 90
symbolic representation 38, 110
sympathy 18, 26
synthesis 52–3, 55–9, 108, 114, 126, 132, 138, 144, 146, 148–9, 156, 159

temporal 91, 99, 101, 105, 112, 114–15, 128–9
textual-constraint objection 11, 17, 23, 25, 37
thematic statement 2–6, 14–18, 38, 121, 165
thematic truthfulness 3–4
third-personal perspective 88, 91
thought experiment 2, 32–5 *See* philosophical knowledge
Tinguely, Joseph 158
Tolstoy, Leo 80
tone 97, 101, 103–5, 108, 115
true belief 13–15, 19, 40, 93, 194
truth 3–5, 12–13, 35, 41, 47, 49, 59, 66, 80, 81, 136, 145, 165, 167, 177

ugliness 125, 129, 132–5, 138, 180
understanding
    axiological understanding 5, 35–9, 166
    explanatory understanding 40, 43
    objectual 5–9, 18–19, 21–2, 38–45, 51–2, 55–7, 59–69, 71–2, 78–9, 85–6, 92, 97, 99, 101–2, 116–17, 121–2, 126–9, 135–6, 138–50, 152–3, 155–62, 166–9, 171–2, 177
unification 149, 156–7
unity 51, 90, 118, 139–40, 142, 144, 146, 148, 152, 155–6, 160

Zagzebski, Linda 40–3

www.ingramcontent.com/pod-product-compliance
Lightning Source LLC
Chambersburg PA
CBHW052117300426
44116CB00010B/1701